TCC South

Crowley Learning Center

·EXPLORING·
SCIENCE AND MEDICAL DISCOVERIES

Astronomy

EXPLORING

SCIENCE AND MEDICAL DISCOVERIES

Astronomy

Clay Farris Naff, *Book Editor*

Bonnie Szumski, *Publisher*
Helen Cothran, *Managing Editor*
David M. Haugen, *Series Editor*

GREENHAVEN PRESS
An imprint of Thomson Gale, a part of The Thomson Corporation

Detroit • New York • San Francisco • San Diego • New Haven, Conn.
Waterville, Maine • London • Munich

Cover credit: © Frank Zullo/Photo Researchers, Inc.
Chris Jouan, 109
Library of Congress, 60
NASA Kennedy Space Center, 15
NASA Marshall Space Flight Center, 175

LIBRARY OF CONGRESS CATALOGING-IN-PUBLICATION DATA

Astronomy / Clay Farris Naff, book editor.
    p. cm. — (Exploring science and medical discoveries)
    Includes bibliographical references and index.
    ISBN 0-7377-2815-9 (lib. bdg. : alk. paper)
    1. Astronomy—History. I. Naff, Clay Farris. II. Series.
    QB15.A88   2006
    520'.9—dc22

2005051160

CONTENTS

ter of the solar system, and all the planets, including Earth, circle it.

2. The Advent of the Telescope

No one knows exactly when the telescope was invented, but its use in astronomy is believed to have started with the Italian scientist Galileo Galilei in 1609.

3. The Planets Travel Around the Sun

In an imagined dialogue, a famous Italian scientist uses logic to try to persuade a disbeliever that the earth orbits around the sun.

4. Newton Formulates Laws of Gravity and Motion

Sir Isaac Newton's discoveries about the universal laws of motion and gravity helped explain the principles that govern orbits.

5. Einstein's Theory of Relativity Advances Astronomy

The director of a planetarium describes how Einstein's 1905 theory of relativity was proven correct during an eclipse of the sun. The theory, which posits that gravity bends light, has provided astronomers with a means of learning about objects too distant from Earth to be observed using conventional telescopes.

6. Hubble Discovers the Expanding Universe

In 1929 astronomer Edwin Hubble analyzed images of galaxies and discovered that they are rushing away from one another. This discovery led to the realization that the universe is expanding.

Chapter 3: Astronomy in the Age of Space Exploration

in the discovery of dozens of massive planets outside the solar system, but none appear likely to harbor life.

Chapter 4: Modern Controversies in Astronomy

Most great science and medical discoveries emerge slowly from the work of generations of scientists. In their laboratories, far removed from the public eye, scientists seek cures for human diseases, explore more efficient methods to feed the world's hungry, and develop technologies to improve quality of life. A scientist, trained in the scientific method, may spend his or her entire career doggedly pursuing a goal such as a cure for cancer or the invention of a new drug. In the pursuit of these goals, most scientists are single-minded, rarely thinking about the moral and ethical issues that might arise once their new ideas come into the public view. Indeed, it could be argued that scientific inquiry requires just that type of objectivity.

Moral and ethical assessments of scientific discoveries are quite often made by the unscientific—the public—sometimes for good, sometimes for ill. When a discovery is unveiled to society, intense scrutiny often ensues. The media report on it, politicians debate how it should be regulated, ethicists analyze its impact on society, authors vilify or glorify it, and the public struggles to determine whether the new development is friend or foe. Even without fully understanding the discovery or its potential impact, the public will often demand that further inquiry be stopped. Despite such negative reactions, however, scientists rarely quit their pursuits; they merely find ways around the roadblocks.

Embryonic stem cell research, for example, illustrates this tension between science and public response. Scientists engage in embryonic stem cell research in an effort to treat diseases such as Parkinson's and diabetes that are the result of cellular dysfunction. Embryonic stem cells can be derived from early-stage embryos, or blastocysts, and coaxed to form any kind of human cell or tissue. These can then be used to replace damaged or diseased tissues in those suffering from intractable diseases. Many researchers believe that the use of embryonic stem cells to treat human diseases promises to be one of the most important advancements in medicine.

However, embryonic stem cell experiments are highly controversial in the public sphere. At the center of the tumult is the fact that in order to create embryonic stem cell lines, human embryos must be destroyed. Blastocysts often come from fertilized eggs that are left over from fertility treatments. Critics argue that since blastocysts have the capacity to grow into human beings, they should be granted the full range of rights given to all humans, including the right not to be experimented on. These analysts contend, therefore, that destroying embryos is unethical. This argument received attention in the highest office of the United States. President George W. Bush agreed with the critics, and in August 2001 he announced that scientists using federal funds to conduct embryonic stem cell research would be restricted to using existing cell lines. He argued that limiting research to existing lines would prevent any new blastocysts from being destroyed for research.

Scientists have criticized Bush's decision, saying that restricting research to existing cell lines severely limits the number and types of experiments that can be conducted. Despite this considerable roadblock, however, scientists quickly set to work trying to figure out a way to continue their valuable research. Unsurprisingly, as the regulatory environment in the United States becomes restrictive, advancements occur elsewhere. A good example concerns the latest development in the field. On February 12, 2004, professor Hwang Yoon-Young of Hanyang University in Seoul, South Korea, announced that he was the first to clone a human embryo and then extract embryonic stem cells from it. Hwang's research means that scientists may no longer need to use blastocysts to perform stem cell research. Scientists around the world extol the achievement as a major step in treating human diseases.

The debate surrounding embryonic stem cell research illustrates the moral and ethical pressure that the public brings to bear on the scientific community. However, while nonexperts often criticize scientists for not considering the potential negative impact of their work, ironically the public's reaction against such discoveries can produce harmful results as well. For example, although the outcry against embryonic stem cell research in the United States has resulted in fewer embryos being destroyed, those with Parkinson's, such as actor Michael J. Fox, have argued that prohibiting the development of new stem cell lines ultimately will prevent a timely cure for the disease that is killing Fox and thousands of others.

Greenhaven Press's Exploring Science and Medical Discover-

ies series explores the public uproar that often follows the disclo-
sure of scientific advances in fields such as stem cell research.
Each anthology traces the history of one major scientific or medical discovery, investigates society's reaction to the breakthrough,
and explores potential new applications and avenues of research.
Primary sources provide readers with eyewitness accounts of crucial moments in the discovery process, and secondary sources offer historical perspectives on the scientific achievement and society's reaction to it. Volumes also contain useful research tools,
including an introductory essay providing important context, and
an annotated table of contents enabling students to quickly locate
selections of interest. A thorough index helps readers locate content easily, a detailed chronology helps students trace the history
of the discovery, and an extensive bibliography guides readers interested in pursuing further research.

Greenhaven Press's Exploring Science and Medical Discoveries series provides readers with inspiring accounts of how generations of scientists made the world's great discoveries possible and
investigates the tremendous impact those innovations have had on
the world.

Astronomy's Bright Future in Space

I n 1909 Edward C. Pickering, director of the Harvard College Observatory and a leading astronomer of his day, was invited to give a commencement address on the future of astronomy. His vision was this: "There will be at least one very large observatory employing one or two hundred assistants, and maintaining three stations. Two of these will be observing stations, one in the western part of the United States . . . the other similarly situated in the southern hemisphere, probably in South Africa." After laying out the details, he added, "These plans may seem to you visionary, and too Utopian for the twentieth century."[1]

It seems never to have occurred to Pickering that anyone might venture out into space or send a telescope aloft. Before the twentieth century came to a close, however, not only were there dozens of major new observatories on Earth but astronauts had been to the moon and back, robotic explorers had landed on Venus and Mars, the Voyager probes were journeying out of the solar system, and the Hubble Space Telescope, orbiting high above the earth's atmosphere, was producing images of celestial objects billions of light years away.

Today, questions about astronomy's future center on whether astronomical research will be conducted primarily by manned or unmanned missions into space. Thanks to *Star Trek* and other science-fictional television programs and films, several generations of Americans have grown up accustomed to the idea that in the future

humans will travel throughout the universe on missions of exploration. Indeed, for this reason, many Americans approve of tax dol-
lars being spent on manned space efforts. However, manned missions are expensive and risky, and many experts argue that science
would be best advanced by investing in unmanned space exploration. Before examining various aspects of the issue, a look back
at the evolution of space exploration may provide some perspective on the forces currently shaping astronomy's future.

Dawn of the Space Age

The year that Pickering gave his address on the future of astronomy,
another American scientist, Robert Goddard, was just beginning to
sketch plans for a liquid-fueled rocket. Ten years later, in 1919, he
published his research on how multistage rockets could be built to
escape the earth's atmosphere. Mocked in the press, Goddard was
forced to carry on his rocket experiments in seclusion. With no government support, he funded his research largely out of his own
pocket. His persistence paid big dividends, for his work in developing rockets was a crucial step in the exploration of space.

Goddard is now widely recognized as the father of modern
rocketry. There were other pioneers in rocketry, however, notably
physicists Hermann Oberth and Wernher von Braun in Germany,
and rocket scientists Konstantin Tsiolkovsky and Sergei Korolev
in the Soviet Union. Beginning in the 1930s German dictator
Adolf Hitler and Soviet dictator Joseph Stalin both ordered secret
rocket-development programs. Since the U.S. government took
little interest in rockets until after the defeat of Germany in World
War II, the Soviet Union (USSR) became the first nation to put a
satellite into orbit. The launch of *Sputnik* on October 4, 1957,
raised the curtain on the age of space exploration.

Cold War Competition

The United States hurriedly stepped up its own plans for space exploration. In October 1958, just a year after the *Sputnik* launch, a
new U.S. government civilian space agency called the National
Aeronautics and Space Administration (NASA) began operations.

Cold War competition made each nation desperate to achieve
firsts in space exploration. Each wanted to show the world that it
was the technological leader. Starting in 1958, both sides hurled un-

manned probes at the moon. Most of these either missed their target or failed to get off the ground. In 1959, however, the Soviet *Luna 3* probe transmitted the first photographs of the far side of the moon. Gradually, the United States caught up, and by 1964 the Ranger series of probes began to send back high-quality images of the moon.

Meanwhile, however, the desire for national prestige and technical mastery drove both space programs in the direction of manned exploration, where the challenges and potential rewards were greatest. Again the Soviet Union got there first with the launch of a rocket carrying cosmonaut Yuri Gagarin on April 12, 1961. Less than a month later, the United States followed with a brief space flight by astronaut Alan Shepard. Soon after, U.S. president John F. Kennedy announced a historic goal for the American space program. Addressing Congress, he declared, "I believe this nation should commit itself to achieving the goal, before this decade is out, of landing a man on the moon and returning him safely to the earth."[2]

To the astonishment of many, the goal was achieved ahead of the deadline. On July 21, 1969, astronaut Neil Armstrong became the first human to walk on the moon. More significantly for as-

Astronaut Buzz Aldrin walks on the surface of the moon in July 1969. The photograph was taken by fellow astronaut Neil Armstrong.

tronomy, Armstrong and his *Apollo 11* crewmates brought back samples of lunar soil. Astronomy thus became something more than mere observation. It became a hands-on science as well.

Ever since the Apollo missions, many of the most exciting developments in astronomy have come from space-based efforts. However, scientists disagree about whether human or mechanical crews provide the best way to carry out such missions.

The Rise of the Unmanned Probes

For many astronomers and planetary scientists, the manned space program has been more of a distraction than a help. These critics believe that unmanned efforts have proven more worthwhile. Unmanned craft have traveled the solar system, exploring planets and other astronomical phenomena, and more recently, space-based telescopes have begun to provide astronomers with unparalleled images of the universe.

The growth of unmanned planetary exploration has been explosive. In 1962 *Mariner 2*, a probe sent by NASA to Venus, became the first spacecraft to travel to another planet and send images back to Earth. Over the next several decades, the American and Soviet space programs sent dozens of spacecraft on unmanned missions to other planets. At first they concentrated on Venus and Mars, the two planets closest to Earth. Occasionally, scientists would send spacecraft to fly by Mercury. After a launch on March 3, 1972, the U.S. spacecraft *Pioneer 10* became the first to explore one of the outer planets. It flew by Jupiter on December 3, 1973. Nearly five years later, twin craft called *Voyager 1* and *Voyager 2* were launched on missions to fly past the gas giants in the outer solar system: Jupiter, Saturn, Uranus, and Neptune. Other probes have made rendezvous with comets, asteroids, and planetary moons. The exploration of Mars has been greatly advanced by the use of robotic rovers, which travel across a variety of Martian landscapes, analyzing samples and beaming back reams of images and data. Among the planets, only Pluto remains unvisited.

Just as significant for astronomers, the unmanned Hubble Space Telescope, launched in 1990, became the first of several orbiting telescopes to provide images unmatched by anything recorded by ground-based observatories. Expressing a widely held view, astrophysicist Joseph Taylor said in 2005 testimony before Congress, "We all love Hubble. It is truly a remarkable instrument."[3]

By contrast, manned exploration appears to have stalled. Since the last lunar landing in 1972, manned missions have been confined to orbiting Earth. The major manned projects—Russia's Soyuz space station, the U.S. space shuttles, and the International Space Station—have all experienced troubles, marked most dramatically by the disintegration of two shuttles in flight. Many in the astronomy and physics community complain that the costly space shuttle and the International Space Station are sucking up funds without scientific payoffs. In a column for the American Physical Society, physicist Bob Park sums up the view of many about the post-Apollo era: "The manned space program has . . . been the antithesis of progress."[4]

Star Trek notwithstanding, some critics think that the moon is the practical limit of manned exploration. They believe that much of the solar system, not to mention other stars, may be off-limits because of the enormous distances involved. For example, as of 2005 the *Voyager 1* spacecraft, after more than twenty-five years of travel, was moving away from Earth at a million miles a day, and yet it remained within the solar system. Scientists estimate that it would take another forty thousand years to reach the nearest star. "For the foreseeable future, robots offer our best scientific reality,"[5] comments Wesley Ward, chief space geologist of the U.S. Geological Survey.

Calls for New Manned Explorations

However, the question of whether astronauts can make significant contributions to astronomy remains open. Advocates for manned missions point out that lack of progress does not imply lack of potential. Some have begun to express frustration with the orbital space program. Testifying before Congress in 2003, Wesley T. Huntress Jr., president of the Planetary Society, which promotes exploration of the solar system, said, "Sooner or later we must have a clear destination for human spaceflight or it will not survive, and America will be much the poorer for it. . . . The country needs the challenge of grander exploration to justify the risk, lift our sights, fuel human dreams, and advance human discovery and knowledge. WE NEED TO GO SOMEWHERE!"[6]

Advocates of manned missions present two main arguments for an expansion of manned exploration. The first is practical. Robotic craft have limitations. Exploration is, by its nature, a venture into the unknown. When trouble arises, robots may lack the creativity and adaptability to solve problems. This has been demonstrated in the numerous failures of unmanned missions. More than half of all craft sent to Mars, for example, have crashed, broken down, or lost communication with controllers back on Earth. Even the highly successful *Opportunity* rover proved helpless when it got stuck in a Martian sandbank. It took Earth-based NASA engineers more than a month of patiently relaying instructions to the rover to turn its wheels one by one before it was able to free itself. By contrast, when *Apollo 13* suffered a crippling explosion and fire on the way to the moon in 1970, astronauts aboard the capsule were able to improvise solutions to the life-threatening challenges they faced.

The second argument for manned exploration concerns vision. Proponents say that the human spirit demands continual exploration of new frontiers. They argue that human destiny is in space. Space enthusiasts regard the manned missions to date as the first steps toward human colonization of the solar system. Aboard the International Space Station, commander Ken Bowersox remarked, "The reason I come up here to space is because I believe . . . that we're laying the foundation for our children, and their children to leave the planet someday."[7]

Early in 2004, President George W. Bush gave an enormous boost to those advocating for manned missions. He announced plans for new manned missions to the moon and beyond that a manned expedition to Mars. "Robotic missions will serve as trailblazers, the advanced guard to the unknown," Bush said. "Yet the human thirst for knowledge ultimately cannot be satisfied by even the most vivid pictures or the most detailed measurements. We need to see and examine and touch for ourselves." Indeed, the president hinted of manned missions of even greater reach in the future: "As our knowledge improves, we'll develop new power generation, propulsion, life support and other systems that can support more distant travels. We do not know where this journey will end. Yet we know this: Human beings are headed into the cosmos."[8]

Some scientists have already made plans for the possibility of human exploration of the galaxy. Designs have been drawn up for interstellar spaceships using enormous solar sails or nuclear explosions as a means of propulsion. In June 2005 a privately funded group attempted to test the responsiveness of a solar sail in space. Unfortunately, the experimental craft was lost in a failed launch, but organizers plan to try again, and NASA has its own solar sail development project. Efforts at interstellar human flight appear certain to continue.

A Balanced Future

Manned space flights are astronomy's flashpoint. Supporters argue that human destiny lies in space and that only manned missions can make the kind of progress that will allow humanity to achieve that destiny. Opponents argue that the costs of manned space flights divert resources from real progress in astronomy and space science. Furthermore, they contend, the risks are so great that it is unethical to put humans in danger when robots can do the job.

However, it appears that a third position is becoming popular. NASA administrator Michael Griffin, who took office in spring of 2005, laid out the position in his first budget testimony before Congress in May of that year. Griffin indicated that the U.S. space agency will pursue both manned and unmanned missions in the future.

> A balanced program of robotic missions will continue to increase our understanding of our home planet and will continue the exploration of the solar system, traveling to the Moon and Mars in anticipation of later human visits, as well as to other destinations such as Mercury, Saturn, Pluto, asteroids, and comets. Observatories will be deployed to search for Earth-like planets and habitable environments around distant stars, and to explore the universe to understand its origin, structure, evolution, and destiny.[9]

Despite the continuing disagreements among advocates and opponents of manned exploration, it appears that Griffin's view reflects the likely future. Perhaps signaling an acceptance of compromise, the American Astronomical Society has declined to take sides on the question of manned or unmanned missions, preferring to emphasize the need to advance scientific exploration of the cosmos by any and all means. Responding to President Bush's announcement of his vision for the civilian space program, the society declared: "The science community is eager to participate in the implementation of the Vision for Space Exploration . . . be it robotic or human."[10]

None of this means that ground-based observations are obsolete. In recent years, astronomers have realized that one of their most critical tasks is to identify asteroids and comets that have a potential to strike the earth. There are far more of these so-called near earth objects (NEOs) than space-based observatories can track. Indeed, even ground-based professional astronomers cannot keep tabs on all the potential threats, so the job has fallen largely to amateurs.

Nevertheless, it seems clear that advances in astronomy will increasingly come from space-based projects. In addition to the plans for new manned missions to the moon and Mars, NASA is planning a replacement for the Hubble telescope sometime after 2010. It will be known as the James Webb Space Telescope. The U.S. space agency also plans to launch an array of space-based telescopes designed to search for Earth-like planets orbiting other stars. Meanwhile, other orbiting platforms such as the Chandra X-Ray Observatory host a variety of astronomical instruments. The European Space Agency has launched a space telescope of its own, the Infrared Space Observatory. If Bush's vision is fulfilled, human explorers will return to the moon by 2020 and will set foot on Mars sometime after that.

If asked to predict the future of astronomy, the current director of the Harvard College Observatory might hesitate to be as specific as Edward Pickering was in 1909. However, one prediction could be made with confidence: The future of astronomy lies largely in space.

Notes

1. Edward C. Pickering, "The Future of Astronomy," commencement address at Case School of Applied Science, Cleveland, May 27, 1909. http://books.jibble.org/1/5/6/3/15636/156368/TheFutureofAstronomy byEdwardCP-0.html.

2. John F. Kennedy, address to Congress, May 25, 1961, quoted in "The Decision to Go to the Moon," Lunar and Planetary Institute. www.lpi. usra.edu/expmoon/decision.html.

3. Joseph H. Taylor Jr., statement to the Committee on Science, U.S. House of Representatives, February 2, 2005. www7.nationalacademies.org/ ocga/testimony/Hubble_Space_Telescope_Repairs_2.asp.

4. Bob Park, "What's New," March 28, 2003. www.bobpark.org/WN03/ wn032803.html.

5. Quoted in Richard Stenger, "Who Should Explore Space, Man or Machine?" CNN, February 18, 2003. www.cnn.com/2003/TECH/space/ 02/18/sprj.colu.space.future.

6. Wesley T. Huntress Jr., "The Future of Human Space Flight," October 16, 2003. www.planetary.org/html/society/wh_testimony.html.

7. Quoted in Stenger, "Who Should Explore Space, Man or Machine?"

8. "President Bush Delivers Remarks on U.S. Space Policy," January 14, 2004. www.nasa.gov/pdf/54868main_bush_trans.pdf.

9. Michael D. Griffin, "Statement of Michael D. Griffin, Administrator, National Aeronautics and Space Administration, Before the Subcommittee on Commerce, Justice, and Science, Committee on Appropriations, United States Senate," May 12, 2005. www.nasa.gov/pdf/115069main_mg_senate_051205.pdf.

10. "American Astronomical Society Statement on the Vision for Space Exploration," undated. www.aas.org/policy/VisionStatement.pdf.

CHAPTER 1

Ancient Skywatchers

Ancient Astronomy and Astrology

By Robert Wilson

Once ancient people settled in one place to grow food rather than hunt or scavenge for it, they began to have more leisure time to engage in intellectual pursuits. Eventually, some began to make systematic astronomical observations. In the following selection Robert Wilson explains that ancient astronomy began to be practiced in Sumeria, Egypt, India, China, and elsewhere as early as 5000 B.C. The ancients studied the heavens for clues about what might befall them—a mystical practice that came to be known as astrology—and for recurring events that could be used in devising calendars to guide planting and harvesting. The Babylonians, Wilson writes, who lived in what is now Iraq, created a calendar and an astrological system whose influence continues to be felt today. They divided the year into twelve months, the circle into 360 degrees, and the sky into a zodiac of constellations. Several other ancient civilizations developed their own distinctive calendars and astrological systems, Wilson writes. The ancient Egyptians, for example, arranged their calendar according to the annual flooding of the Nile. Robert Wilson was a prominent British astrophysicist known for his work on the *International Ultraviolet Explorer* satellite. He died in 2002 at the age of seventy-five.

About 10,000 years ago, the most recent ice age was over, the ice had fully retreated and the resulting warm period led to a spread of forests, vegetation, fish and mammals. The human race, which had hitherto spent its energy and ingenuity on survival—the acquisition of food and provision of warmth—responded to the greatly increased food supply by accelerating the development of tools and the techniques of hunting and gathering of plants. The increased productivity allowed the human race to exploit its greatest gift, a powerful brain, more fully. It embarked on the development of civilization and found that the setting up of organized societies, in the form of tribes or whatever, resulted in greater prosperity and more effective defence. It found that farming the land to produce the crops that it wanted, and the husbandry of animals to produce the meat that it needed, were far more bountiful than gathering vegetation and hunting game which happened to be present naturally. . . .

By *circa* 5000 B.C., food production had reached a level that freed significant time and effort for pursuits beyond those needed for survival alone. The consequent release of human ingenuity caused an acceleration in human progress, with the further development of new tools and techniques, allowing more effective farming (and warfare), the building of cities, the development of language, from spoken to written, the strengthening of social organization and authority, and the creation of wealth. Out of this evolved the first true civilizations: societies with the means and the wish to pursue intellectual, artistic and other creative activities in addition to the most basic needed for survival. The great early civilizations were located in many parts of the globe, usually in the most fertile regions, often watered by great rivers. The first of these was the Sumerian civilization, which was fully established in Mesopotamia by *circa* 3500 B.C., so named because it is the land between two rivers (the Tigris and the Euphrates), and is largely embraced by modern-day Iraq. Others were well established in Egypt by *circa* 3000 B.C., in India (the Indus valley) by *circa* 2500 B.C., in Crete (Minoan) by *circa* 2000 B.C., in China by 1500 B.C., and in Central America (the forerunners of the Incas and the Aztecs) by 1000 B.C.

Driven by Curiosity and Need

Many of the intellectual activities pursued in the early civilizations grew from the innate curiosity of the human race in the natural world in which it existed. This was particularly true of astronomy, where the motion of the Sun, Moon, planets and stars caused excitement and puzzlement. You should realize that life in a modern industrial society is a great impediment to viewing the heavens because of artificial lighting and smog. One of the really beautiful sights in nature is of the clear night sky in the absence of city lights, say on a remote mountain, where the sky has great depth and the Milky Way is a bright lane of light. But this sight was available to everyone in the distant past and its beauty led to the pursuit of astronomical studies in all the early civilizations.

But the early developments in astronomy were fired more by practical and mystical rather than scientific considerations. The development of agriculture required that crops be planted in spring and harvested in autumn; hence the times of the seasons needed to be known. In other words, a calendar was required, and several attempts to establish one were made in the period between 5000 and 1000 B.C. Natural time-periods were available in the day, determined by the rising, setting and rising again of the Sun; the month, determined by the time it took the Moon to pass through all its phases; and the year, determined by the seasons, over which the Sun reached its maximum noon altitude in mid-summer (for the Northern Hemisphere), its lowest in mid-winter and back again in mid-summer. The first and simplest astronomical instrument, the gnomon, was able to give some indication of the time of day and the season of the year; it consisted of a straight vertical rod and is based on the same principle as the modern sundial. In the early morning, its shadow would be long and point roughly westwards; as the day progressed, it would shorten and rotate until, at noon, it would be at its shortest and would point exactly due north; it would then lengthen and rotate until, in the late afternoon, it was pointing roughly eastwards. If the length of the shadow is measured at noon, it is found to be shortest in mid-summer and longest in mid-winter, thereby allowing the seasons to be estimated. Another way to tell the time of year was afforded by the night sky. The stars were fixed and unchanging relative to each other, but appeared to rotate completely over one year, so that there were winter and summer constellations.

The Puzzle of the Planets

Another group of objects in the night sky were the five planets or wandering stars. As bright as the brightest stars, they moved in the same plane (the ecliptic) as the Sun and Moon, but in odd and seemingly unpredictable ways. Three of them (Mars, Jupiter and

Saturn) would advance across the celestial sphere, reverse their motion and advance again; the other two (Mercury and Venus) also moved but were visible only when they were close to the Sun, just after sunset or just before dawn. The puzzle of the planets was to remain unsolved for several millennia and it posed the greatest problem in early cosmologies—the explanation of the motions of the Sun, Moon, planets and stars.

The mystical aspect of studying the heavens, astrology, developed strongly in the early civilizations and soon became the prime driving force. It was believed that the stars and planets controlled human destiny and therefore their study was encouraged as a means of predicting, or explaining, human triumphs and tragedies. Perhaps this is not surprising: if the heavens could say when crops should be planted or harvested, why not when wars should be embarked upon or preparation made for famine or flood? Religious aspects also crept into the interpretation of the heavens, and astronomical studies were often carried out by priests.

Ancient Sumerians

The most ancient civilization, the Sumerian, was based in Mesopotamia, now the southern part of modern-day Iraq. It prospered rapidly and by *circa* 3000 B.C. had developed a written language which was etched into clay tablets. Unlike the other early civilizations of Egypt and China, it has not retained its name, culture and identity over the centuries, but has changed hands frequently through invasions, migrations and wars. The development of building technology allowed for larger and larger human groupings into city states, and one of these, Ur, became the capital of Sumeria in *circa* 2500 B.C. It lay near the junction of the Euphrates and Tigris rivers in the south of Mesopotamia. It is believed to have been the home of Abraham and his semitic tribe, who developed a belief in a single, unseen, all-powerful God, a belief that forms the ancient roots of today's three great monotheistic religions. Abraham was to take his tribe out of Ur *circa* 2100 B.C., when it set off on its wanderings (from which the name Hebrew, or "wanderer", derives), which were to take it to Canaan (Palestine), to Egypt and back to Canaan over a thousand years.

In about 2000 B.C. Ur was conquered by the Elamites, who came from what is now southwest Iran, and a new powerful city-state emerged, Babylon, which was to establish control of Sumeria and extend its empire over the whole of Mesopotamia. Babylon is on the Euphrates about 80 kilometres south of present-day Baghdad. It became very wealthy and very indulgent in the pursuit of pleasure and consumption, giving it a reputation as a magnificent, worldly, wicked city, which still persists today more than two millennia after its demise. Great buildings were constructed, including the Hanging Gardens, one of the seven wonders of the ancient world. The Sumerian cuneiform script was given a syllabic form, thereby greatly increasing its flexibility. Many written tablets still survive which tell us more about the Babylonian Empire than we know about many European countries during the Dark Ages of A.D. 500–1000.

Many activities were encouraged, one of which was a study of the heavens, entirely for astrological purposes, since they believed their destiny could be read in the stars, but this had the important result of producing the first major set of astronomical observations ever undertaken. Over centuries, the positions of the bright stars were established and the motions of the Sun, Moon and planets charted against the background of those fixed stars. All these data were recorded in cuneiform script etched into clay tablets, many of which are still available today; they allowed some astronomical analysis, as well as astrological interpretations. By about 1000 B.C., lunar eclipses could be predicted with reasonable accuracy and the motions of the outer planets had been established over several centuries.

Creation of a Calendar

Studies of the heavens also had religious overtones, and the prime celestial bodies were named after gods, of which the Babylonians had several. But in addition to the religious and astrological applications, there was an astronomical requirement—the prediction of the dates of important events such as festivals and the times of sowing and reaping. This required a calendar and one was constructed which had a year of 12 lunar months, and which started with the vernal equinox in spring, the time for sowing. Each month was defined by the Moon and started at sunset on the evening when the new crescent Moon was visible for the first time. Since a lunar month is not an exact number of days, nor is a solar year an exact number of lunar months, this was a somewhat cumbersome basis for a calendar, as many other civilizations discovered. A lunar month is close to 29.5 days, so, with each month being identified by the first appearance of the new Moon, a month was either 29 or 30 days and a year of 12 lunar months consisted of 354 days. The difference between this and the solar year (365.25 days), which determines the seasons, was accommodated with an extra month about every three years. The Babylonians also introduced the seven-day week in deference to the Sun, Moon and five planets, regarded as representing gods.

The Babylonians were responsible for some of the earliest mathematics and they set up the sexagesimal system of angular measurement which is based around the number six and is still used today. They defined a full circle as 360°, corresponding approximately to the 360 days in their calendar; hence 1° corresponds approximately to the angular movement of the Earth during one day in its orbit around the Sun, although the Babylonians did not look at it in that way. They also set up a simple form of algebraic geometry which had many practical uses and formed a basis for surveying.

Invention of the Zodiac

The Babylonians also initiated the concept of the zodiac, the band in the celestial sphere through which the Sun moves and completes a full revolution in one year. The Moon and planets also move in the same band (but in totally different ways) and we now know that it represents the ecliptic plane in which the planets revolve about the Sun, and the Moon about the Earth. The Babylonians divided it into 12 equal zones of 30° corresponding to the 12 lunar months of their calendar and then assigned each one to the nearest constellation. Constellations were regions of the sky whose stars showed a pattern which could be likened to known people, animals or objects. All the early civilizations identified their own constellations and the 12 signs of the zodiac evolved slowly from ancient roots. The ones that are in use today are those which were recorded by the ancient Greeks; the sequence starts with Aries (the Ram), moves to Taurus (the Bull), then to Gemini (the Twins), to Cancer (the Crab), on to Leo (the Lion), on to Virgo (the Virgin), then to Libra (the Balance or Scales), on to Scorpio (the Scorpion), then to Sagittarius (the Archer), reaching Capricorn (the Goat), then on to Aquarius (the Water Bearer) and to Pisces (the Fishes), which then joins up with Aries to complete the cycle.

During each year, the dates that correspond to each sign of the zodiac are determined entirely by the constellation in which the Sun is located in its migration around the zodiacal plane. The zodiacal year is defined by the time it takes the Sun to complete its full cycle; it is now called the sidereal year, because it is determined by the Sun's movement against the background of fixed stars. In today's terms, it is precisely the time it takes the Earth to complete one full revolution about the Sun. In developing a calendar, the Babylonians knew that a year determined the cycle of seasons, which we now know are caused by the tilt of the Earth's axis of rotation by 23.5° to the plane of its orbit around the Sun. In summer the axis is tilted towards the Sun, causing greater heating because of the longer period of daylight and the greater solar radiation flux per unit area because of the angle of the Earth's surface being less inclined to the direction of the Sun's rays. Midsummer is defined by the Earth's axis being tilted exactly towards the Sun (the summer solstice), and mid-winter when it is tilted exactly away from the Sun (the winter solstice). (This is the situation for the Northern Hemisphere; the inverse is the case in the Southern Hemisphere.) Spring is defined by the vernal equinox when the Earth's axis is tilted exactly orthogonal to the Earth-Sun line, so that the Earth is equally illuminated from pole to pole and periods of day and night are equal everywhere; similarly, autumn is defined by the autumnal equinox. Hence, a year can also be determined by measuring the direction of the Earth's inclination relative to the Sun, say from vernal equinox to vernal equinox. Such a year, called a solar year, is directly linked to the seasons and is therefore the most important basis for a calendar, the greatest practical value of which is to predict the occurence of spring, summer, autumn and winter or, more importantly in Egypt, the flooding of the Nile and, in India, the monsoon, Because of this, our modern calendar . . . is based on the solar year.

Astrological Error

At this point, you may think that the two years I have described—the zodiacal, or sidereal year and the solar year—are identical and simply represent two different ways of measuring the same thing, but this is not so; there is a very small difference between them. The difference is so small that it was unknown to the Babylonians and was not detected until nearly 2000 years after they started studies of the zodiac and their calendar. It is so small that it has a negligible effect on everyday life and, even today, it is not well known outside astronomical circles. However, there is one area of modern activity that should be affected by the small difference between solar and sidereal years but which seems to be largely ignorant of it.

As already said, much of early astronomy was driven by astrology, the belief that human destiny can be foretold in the stars. The zodiac became one of its main tools and, since it is still used today, as can be seen from its presence in the columns of many newspapers, it is appropriate that I bring its story up to date. The simple proposition is to assign to each individual person that zodiacal sign that the Sun was located in when he or she was born. Destinies could then be read at any one time from the positions of the Moon or planets as they moved through the different signs of the zodiac. However, the direction of the Earth's axis is not fixed but, because of its rotation, precesses in a way similar to that of a spinning top; it closely maintains a tilt of $23.5°$ to its orbital plane but rotates very slowly about it at a rate that will see complete revolution in 25,800 years. Because of this the constellations move slowly westwards along the plane of the ecliptic with respect to the calendar. This is because the calendar is based on the solar year, which, because of the precession of the Earth's axis, is about 20 minutes shorter than the sidereal or zodiacal year. Without precession, these two years would be identical and the zodiacal constellations would not drift through the calendar. Precession was discovered in the second century before Christ by Hipparchus, the most brilliant observational astronomer of ancient Greece. . . .

The Babylonians had adopted the vernal equinox, marking spring, as the start of their year and the zodiac. At the time of the ancient Greeks, the vernal equinox lay in Aries, so this constellation was labelled as the first sign of the zodiac and the vernal equinox is still referred to as the first point of Aries. But when the Babylonians first set up their zodiac, the vernal equinox lay in Taurus; this was therefore the first sign of their zodiac and it accounts for the Babylonian description of Taurus as not only "The Bull of Heaven" but also "The Bull in Front". If our present series of zodiacal signs had been adopted from the earlier Babylonian records rather than the later Greek ones, the first sign of the zodiac would have been Taurus, not Aries, and the astrological discrepancy I am describing would have been even greater. Today the vernal equinox, which occurs on 21 March in our calendar, lies in Pisces; hence, most people who are assigned the astrological sign of Aries should, astronomically, have the sign Pisces. This drift will continue and, during the next millennium, the vernal equinox will enter Aquarius, presumably heralding the "dawn of Aquarius" Hence, if anyone wished to know which constellation the Sun was in on any particular date, past or future, the appropriate astronomical tables would be needed or, more simply, a sidereal rather than a solar year used in which the constellations will not drift. A more simple but approximate calculation can be made from the fact that the constellations drift through our calendar by one zodiacal sign every 2150 years and that the vernal equinox lay in Aries during the great Greek period of a few, say four, centuries before Christ. As of today, this places the vernal equinox in the western part of Pisces. . . .

Other Ancient Civilizations

The other great early civilizations, such as those in Egypt, India and China, also conducted astronomical studies which were driven by practical, astrological and religious motives. In China there is evidence that astronomical observations were well under way as long ago as *circa* 2000 B.C. The length of the solar year and the lunar month had been determined with good accuracy, allowing a calendar to be developed which could establish the times of festivals and so on. Constellations were identified with emperors, and stars were ranked in status according to their brightness. They had a very simple cosmology: the Universe was the rotating sphere of fixed stars in which the poles were the exalted regions. Astrology was dominant, but little attention was given to the motion of the planets, an important basis of many other astrologies. Instead, changes such as the appearance of a new star or comet were believed to herald calamities on Earth. One of the most dramatic of such celestial changes is a total solar eclipse, and legend has it that two astrologers were executed because they failed to predict one that occurred in 2137 B.C. In India, astronomy developed a little later than in China but in a similar fashion. The solar year was measured, a calendar was constructed and the major celestial objects were named after gods and goddesses; astrology was the driving force in these endeavours.

Of course, our knowledge of such ancient activities depends entirely on the evidence handed down to us. The most effective evidence is in written manuscripts, but we must recognize that many of the earliest writings were based on word of mouth transmission through many generations and were therefore written much later (sometimes centuries) after the events they record. Other astronomical developments, of which we do not have a written record, certainly occurred in other parts of the ancient world, and of these we know little or nothing. One example is the megalithic monuments that are found in many countries, especially the British Isles, and which date as far back as the third millennium B.C. These represent a considerable engineering feat (some of the stones are as heavy as about 50 tonnes) and their layout, which shows considerable trigonometric knowledge, is clearly dictated by astronomical principles. The most notable of these is Stonehenge in the south of England.

In the above account of astronomical studies in the ancient world, I have given the most detailed description to those in Babylon because they were the most extensive and had the soundest mathematical base. But one other civilization, Egypt, was the first to construct a calendar and, with subsequent developments, provided the basis of the calendar we use today. . . .

Mistaking Coincidence for Cause

It so happens that, when the Egyptians were developing their calendar, the flooding of the Nile was heralded by the heliacal rising of the brightest star in the sky, Sirius (Sothis to the Egyptians), which occurred in mid-June as seen from Egypt. The heliacal rising occurs in the eastern sky just before dawn when the star reappears after having been invisible for about 70 days because of being in daylight. This remarkable event—the reappearance of the brightest star in the sky just before the most important event in Egyptian life, the flooding of the Nile—greatly strengthened the belief in astrology. . . .

Out of [this and other] beliefs emerged an explanation of the annual flooding of the Nile on the heliacal rising of Sirius, which was an imaginative and romantic combination of religion, mysticism and astronomy. The heliacal rising of Orion occurs before that of Sirius, so, star by star, Osiris [god of fertility] was revealed by this most magnificent of constellations straddling the celestial equator. Then, on the later heliacal rising of Sirius, Isis [consort of Osiris] was also revealed and, seeing her husband, she was immediately saddened and went into mourning for his death. Being a very loving and faithful wife, her sorrow was such as to cause the tears she shed to be so profuse that they inundated the valley of the Nile.

The Egyptians were the first to recognize that the naturally available time intervals of a day, month and year were not commensurable and they were therefore the first to appreciate the difficulty this posed for the construction of a calendar. They took the first major step in developing their own calendar by accurately determining the length of a year as 365.25 days by timing the occurrence of the equinoxes over a long period. They were therefore measuring a solar year, and since this determines the seasons, it is the best basis for a calendar and is the basis of the one we use today. To avoid any confusion, it should be stressed that the small difference between the solar year and the sidereal year, discussed above in connection with the zodiac and caused by the precession of the Earth's axis, has no effect on this story of the development of the calendar. The difference can be detected only by reference to the fixed stars, as was the case with the zodiac, and would ultimately have been revealed to the Egyptians because precession causes the heliacal rising of Sirius to occur later as time progresses. Today, some five millennia later, the rising is no longer in mid-June but in August and it can no longer be regarded as the herald of the flood. . . .

Roman Calendar

In the early days of its empire, Rome adopted a calendar essentially based on that of Babylon, nearly two millennia before. There were 12 months determined by the first appearance of the new Moon, and therefore of 29 or 30 days' duration, giving a year of 354 days. Since this rapidly got out of step with the seasons, it was corrected, as in Babylon, by the insertion of an additional month every three years or so. Like the Babylonians, the Romans started their year at the vernal equinox and named their months by number. Some of these still survive in our calendar today, where September, October, November and December are derived from the Latin words for seven, eight, nine and ten. Of course they do not correspond to those numbers in our calendar today because the first month of the old Roman calendar coincided approximately with our March. . . .

In order to avoid any future adjustments to the calendar caused by the solar year being slightly less than 365.25 days, a fine tuning of the leap-year rule was introduced in which century years not divisible by 400 would not have an extra day added, even though they meet the normal four-year rule. Hence, 1700, 1800 and 1900 were not leap years but 2000 will be. The Gregorian calendar was immediately adopted by all Roman Catholic European countries, but in those early post-reformation years, the Protestant states of northern Europe and the orthodox Christian countries of eastern Europe did not follow suit. Britain waited until 1752, when 11 days had to be deleted from the Julian calendar, and Russia did not change until the twentieth century, when 13 days had to be deleted. This explains why the Bolshevik revolution of 1917 is called the "October" revolution; it fell in that month on the 25th in the Julian calendar although it now falls on 7 November in the Gregorian calendar.

The Ancient Greek Astronomers

By James Trefil

The ancient Greeks had no celestial maps, telescopes, or other such instruments, but they had a tradition of scientific thinking. In the selection that follows, astronomer James Trefil describes some of the remarkable discoveries made by the Greeks, along with some of their errors. Trefil notes that several Greek astronomers came out against the common perception that Earth was flat. Most notable among these was Claudius Ptolemy. Writing in the second century, Ptolemy argued that Earth was a sphere. Among his supporting arguments was the observation that as a ship approaches the shore, mountains appear to rise from beneath the horizon. Ptolemy also argued, wrongly, that Earth was the unmoving center of the universe. His theory was generally accepted until Copernicus overturned it many centuries later.

Although Ptolemy is perhaps best known, other Greeks contributed to astronomy as well. An ingenious Greek named Eratosthenes not only showed that Earth is round but made a reasonably accurate estimate of its circumference. Hipparchus, Trefil writes, contributed an understanding of Earth's precession, or wobble, around its axis, which accounts for the gradual shift in the position of constellations. Several ancient Greeks argued for a heliocentric system. Most notable among them was Aristarchus, who argued that the sun was located at the center of the solar system. He was right, but his claim seemed unbelievable at the time. Astronomer and science writer James Trefil is the Clarence J. Robinson Professor of Physics at George Mason University.

James Trefil, "Rounding the Earth," *Astronomy,* vol. 28, August 2000, p. 40. Copyright © 2000 by Astronomy Magazine, Kalmbach Publishing Co. Reproduced by permission.

Everyone who lived before Christopher Columbus thought Earth was flat, right? Actually, Greek scientists knew Earth was a sphere thousands of years before Columbus first sailed the Atlantic. That knowledge was incorporated into the book that served as the standard astronomy text in medieval universities. Columbus didn't have to convince educated men and women that he could reach the East by sailing west. His task was more practical. He had to convince them that Earth's circumference was small enough to allow him to reach China before his ships would run out of food and drinking water.

Universal belief in a flat Earth in Columbus's day is a myth—one example of a modern tendency to assume those who came before us were less able to cope with the universe. The cumulative nature of science is partly to blame. Each discovery raises new questions, which scientists answer later. Modern college students know more about physics than did Isaac Newton. Of course, that doesn't make the students geniuses. Indeed, the more I discover what ancient scientists did with primitive instruments, the more impressed I become.

Take the shape of Earth. Second-century Egyptian astronomer Claudius Ptolemy (evidence suggests he died in A.D. 141 or 151 at age 78) compiled the work of astronomers dating back 400 years. He called the book *The Mathematical Composition*, but to later generations it became known as the *Almagest*, a word put together by Arab astronomers from their article Al and the Greek megiste, or great. This book was the encyclopedia of astronomy until the Renaissance—a shelf life that modern publishers can only dream about. (As a geographer Ptolemy also produced a collection of maps in *Geography*, which Columbus read and concluded that it must be possible to reach India by sailing west.)

The Roundness of the Earth

A few pages into Book I of the *Almagest* we find a section titled "That Also Earth, Taken as a Whole, Is Sensibly Spherical." In a single densely worded page, Ptolemy gives a variety of arguments to support the proposition that Earth is round, including these:

The sun, moon, and stars do not rise at the same time everywhere on Earth. Events such as lunar eclipses that take place at a specific time as seen from any vantage point occur at earlier local times in the west than in the east. If Earth were flat, the event would occur at the same local time at two distant locations.

The difference in local time between these events is proportional to the distance between them, as you would expect for points located on the surface of a sphere.

As an observer travels north, southern constellations disappear from view and northern constellations move higher in the sky.

When a ship sails toward shore, a mountain may appear to rise from the sea, as would be expected if its lower parts were hidden by Earth's curvature.

None of these observations requires sophisticated equipment— just an ability to put together everyday observations. Of course, this kind of reliance on common sense and everyday observation can get you into trouble. For example, the same sorts of arguments a few pages later in Ptolemy's book bolster the notion that our planet is the stationary center of the universe. In this section, Ptolemy marshals the best science of his day to support this conclusion.

Aristotle had taught (wrongly) that heavy objects fall toward the center of the universe faster than light ones. Ptolemy argued that if Earth were not at the center of the universe, it would be falling toward the center. Furthermore, because of its great mass, Earth would fall faster than anything would on it, leaving human beings, dogs, and the pyramids floating in space.

I don't mention these points as criticism of Ptolemy. After all, many centuries would pass before the genius of scientists such as Galileo, Isaac Newton, and Robert Hooke worked out the subtle connections between gravity and centrifugal force that, in the end, answered Ptolemy's objections.

Estimating the Size of the Earth

Actually, Ptolemy wasn't the first to argue that Earth was round. The first were followers of the mathematician Pythagoras around 500 B.C. They appear to have argued that Earth was spherical and rotated on its axis, although they did not apply Ptolemy's rigor to the subject. Furthermore, about 350 years before Ptolemy set down his ideas, a fellow Alexandrian scientist trained in the Greek tradition—Eratosthenes—not only accepted the spherical shape of Earth, but also devised a way to measure its circumference.

Eratosthenes was born in the town of Syene (now Aswan, Egypt), which is near the Tropic of Cancer. To get Earth's radius, Eratosthenes dug a deep well and made two measurements. At noon on the summer solstice, when the sun was directly above Syene, sunlight would illuminate the bottom of a well there. Also at noon on the same day, he measured the length of a shadow cast by a pointer fixed in a hemispherical bowl in Alexandria. Knowing that the direction of the sun's rays are the same at Syene and Alexandria, and noting that the length of the shadow in Alexandria stretched over 1/25 of the hemisphere (and hence over 1/50 of a sphere), he concluded that the distance between the two towns was 1/50 of the circumference of Earth.

So far, so good. Modern astronomers confirm that the angle measurement is about right. To convert it to an estimate of Earth's radius, however, Eratosthenes had to know the distance between Syene and Alexandria. He did not have the kind of trigonometry that would make a modern survey possible, so he had to rely on a less accurate estimate. The rulers of Alexandria employed a core of "pacers"—men who counted their steps as they trekked from town to town. From their reports, Eratosthenes concluded that the two towns were separated by 5,000 stades, a Greek measurement from which we derive stadium. (One stade is about 600 feet.) Eratosthenes reported Earth's circumference to be 252,000 stades (he apparently threw in the extra 2,000 stades so that the final number would be divisible by 60). A judgment of the accuracy of this figure depends on which of the ancient stades Eratosthenes actually used—a subject of scholarly debate. His most likely estimate of Earth's circumference was about 29,000 miles—close enough to the current value of about 25,000 miles to command our admiration.

Of course, there were problems with Eratosthenes's method. Syene is not due south of Alexandria (as it would have to be for his method to work). Furthermore, the distance between the two is about 4,530 stades, rather than 5,000. Nevertheless, Eratosthenes's feat demonstrates the power of clear reasoning.

Contributions of Hipparchus

A similar lesson can be drawn from the work of another Greek astronomer, Hipparchus. Born in a Greek city in what is now western Turkey, he did most of his astronomical work on the island of Rhodes in the Aegean Sea. (He died there sometime around 127 B.C.) He was one of the greatest, if not the greatest, of the ancient astronomers. Until recently, some scholars argued that Ptolemy had merely copied Hipparchus's writings to create the *Almagest.* It is disappointing that we know so little about him or his life. The only surviving anecdote (of doubtful veracity) is that on one clear evening he once wore a cloak to a theater performance in Rhodes because he had predicted a storm.

Hipparchus's contributions to astronomy were prodigious. He was the first to record the positions of stars in the sky. (He apparently claimed he did so to be able to tell time at night—another first.) To honor Hipparchus's work, the European Space Agency named a satellite for him that was used to determine the positions of stars to unprecedented accuracy.

Hipparchus also compiled eclipse records from Babylonian astronomers and used them with other measurements, according to Pliny the Elder, to predict eclipses hundreds of years in the future (modern scholars sometimes dispute this claim). But his greatest fame rests on two discoveries: the so-called precession of the equinoxes and his partially successful attempt to determine distances between objects in the solar system and their sizes.

The precession of the equinoxes is understandable if we think of Earth as revolving around an axis that tilts about 23 Degrees from the plane of its orbit, called the ecliptic. Furthermore, the axis of rotation isn't fixed in space but wobbles like a spinning top. In this way, Earth completes one wobble in about 26,000 years. This precession makes true north move away from the North Star. In 13,000 years, for example, our North Star won't be Polaris. It will be Vega. In 26,000 years, Polaris will again be the North Star. This precessionary wobble has other consequences. For example, the position of the sun with respect to the stars at the spring and fall equinoxes also appears to move around the sky every 26,000 years. It was this subtle motion that Hipparchus detected with his careful measurements and historical records.

Distances to the Sun and Moon

Hipparchus achieved his greatest impact with his estimates of the sizes of the sun and moon and the distances that separate them from Earth. Estimating these distances continued through millennia and continues in the quest to measure the basic dimensions of the universe.

On March 14, 190 B.C., an eclipse of the sun occurred in the Middle East. Comparing the angle of sight to the moon from a spot near the present location of Istanbul with the angle measured at Alexandria and using some simple geometry, Hipparchus concluded that the distance to the moon on that day was about 71 times the radius of Earth. (That's not bad—the currently accepted value of the Earth-moon distance is about 60 times Earth's radius.) He was less successful in estimating the distance to the sun—his value was 490 Earth radii or 194,000 miles, while the actual distance is roughly 93 million miles.

There is a legend that Hipparchus's ability to predict eclipses was foreshadowed by one of his predecessors, Thales of Miletus. Called the "Father of Science" Thales was the first to apply the questioning, naturalistic mode of thought we identify with modern science to a wide range of topics. His students, for example, devised the familiar Earth, fire, air, and water system of explaining the properties of materials. We know little about Thales. According to legend, he fell into a ditch because he was looking up at the stars. He supposedly stopped a battle between the Medes and the Lyrians with his prediction of an eclipse in 584 B.C. Most modern scholars discount the eclipse-prediction story because the techniques needed to predict eclipses were developed hundreds of years later.

A heretical, minority opinion united a few ancient astronomers: Earth is not at the center of the cosmos but instead orbits the sun with the other planets. Unfortunately, the documents that support these theories vanished like the people who produced them. All we have are comments made later by writers whose works survive. The notion of a sun-centered system seems to have started with the followers of Pythagoras. He argued that since fire was the purest element, it should be located at the position of pride, in the center of things. Some versions of this theory were truly bizarre— in one, a flat Earth rotated in orbit around the central fire in such a way that the uninhabited side was illuminated. In this creative picture, the sun was apparently a secondary fire that also orbited the center.

Aristarchus on Heliocentricity

But the scientist universally credited by ancient authors with producing a sun-centered theory was Aristarchus of Samos (310–230 B.C.). The only surviving work of Aristarchus, a book called *On the Sizes and Distances of the Sun and Moon*, shows no hint of heliocentrism, so his theory may have been developed in a manuscript that was lost. Archimedes writes in his *The Sand Reckoner:* "[Aristarchus's] hypotheses are that the fixed stars and the sun are stationary, and that Earth is borne in a circular orbit around the sun."

This sounds convincing to me. But arguments like those advanced by Ptolemy weighed heavily against this viewpoint in the ancient world, as did the religious conviction that the home of humanity had to occupy the central spot in the universe. Among astronomers, another argument was important: If Earth moves in a circle, the stars should appear to move in the sky from one season to the next. This so-called stellar parallax is easily seen with modem telescopes but could not be seen with instruments available in antiquity.

Aristarchus's argument that the stars were too far away for us to see parallax, though it ultimately proved correct, was rejected as special pleading by his contemporaries. They were right to question it. There was precious little evidence to support the sun-centered universe and many reasons to reject it. Relying on someone to introduce a new and totally unsupported idea to save the theory was no more convincing to ancient astronomers than it would be to astronomers today. Scientists see these kinds of arguments all the time and they almost always have an aura of grasping at straws. The chance that any one of them may prove to be right centuries down the road is pretty slim. So in the end, I would be reluctant to include a heliocentric universe in my list of the accomplishments of ancient astronomers.

However, as we have seen, the list is impressive without it. But contemplating that list brings us to another, and potentially more interesting, question: Why, given all this knowledge, did astronomy need the Renaissance to be reborn? Why, in other words, was there a 1,400-year gap between Ptolemy and Copernicus?

On the Way to Copernicus

The short answer to this question is that there wasn't—that the notion of a gap results from too narrow a view of history. While Europe was sinking into its Dark Ages, the compilation of astronomical knowledge in Ptolemy's book became the basic working tool of Arabic astronomers. They spent centuries refining their observations and producing commentaries on the *Almagest.* When Gerard of Cremona translated the book from Arabic to Latin in 1175, the main corpus of ancient astronomy once again became available in the West and, along with the works of Aristotle and Plato, was quickly incorporated into universities' curricula.

The Polish cleric and political leader Nicolas Copernicus (1473–1543), the person generally reckoned to be the founder of modern astronomy, studied the *Almagest* closely. It was the custom at the time of Copernicus for men like him to have an intellectual hobby. He asked a simple question: Is it possible to explain the data compiled by Ptolemy with a system in which the sun, rather than Earth, is at the center of the universe?

Most of the data he used was from the tables in the *Almagest*, supplemented by a few observations of his own. The mathematics at Ptolemy's disposal and the type of reasoning he used would also have been available to any scholar after him. Copernicus' great achievement was to take the then-outlandish notion of heliocentricity and develop it to the point where it could be compared, exact prediction to exact prediction, to Ptolemy's Earth-centered universe.

Like all giants of science, Copernicus relied on the work of those who came before him, including Ptolemy. For example, in order to explain the motion of planets, Ptolemy said they were attached to small crystal spheres called epicycles. These epicycles rolled around on larger crystal spheres in the sky. Copernicus's model also had the planets, including Earth, moving around on epicycles. When critics objected to a rotating Earth, Copernicus was quick to point out that rotation involved motion within a circle, the perfect geometrical form. This was a classic and medieval argument, but hardly a scientific one.

Copernicus took the universe bequeathed him by the ancients and transformed it by asking a fundamental question. He didn't give us a modern understanding of the universe, but he placed us on the road leading toward it. His single insight changed forever the way we viewed the universe and our place in it. What better tribute to the work of the ancient astronomers?

The Earth Is the Center of the Universe

By Claudius Ptolemy

The astronomer with the longest-lasting influence on the Western world's understanding of the universe was unquestionably Claudius Ptolemy (ca. 85–150). Ptolemy devised a geocentric theory that withstood all challenges for some fourteen centuries. To be sure, Ptolemy's ideas endured in part because they coincided with those of the widely revered philosopher Aristotle, who centuries earlier had offered a geocentric view of the universe. However, Ptolemy offered powerful mathematical and logical arguments of his own, which proved exceedingly difficult to refute.

In the following selection from his masterwork, *Almagest*, Ptolemy explains why he believes that the stars and planets revolve around an unmoving Earth in perfectly circular orbits and suborbits. He observes, for example, that the stars never appear to move closer or farther from Earth or from one another. This could only be so, he argues, if they were fixed in a sphere rotating on an axis with Earth at its center. It is now known that the then-unimaginable distances of the stars accounts for their apparent lack of movement as Earth orbits the sun. Not all of Ptolemy's ideas were wrong. For example, he proposed that Earth is round. Despite some missteps, such as his argument against Earth's rotation, Ptolemy gave the world a system for tracking the movement of the planets that, while cumbersome, worked well enough for astronomers to use for many centuries.

Ptolemy was born into a Greek family whose members enjoyed the privileges of Roman citizenship during a time when Rome ruled Alexandria, the Egyptian coastal city where he spent most of his life. There, he made astronomical observations and wrote his treatise,

Claudius Ptolemy, "Translation of the Almagest," *Ptolemy's Almagest,* translated and annotated by G.J. Toomer. London: Duckworth, 1984. Copyright © 1984 by G.J. Toomer. All rights reserved. Reproduced by permission of Gerald Duckworth & Co. Ltd.

which after the fall of the Roman Empire survived only among Arab astronomers. Europeans translated it from Arabic beginning in the twelfth century. The work's Arabic name, *Almagest*, stuck.

n the treatise which we propose . . . the first order of business is to grasp the relationship of the earth taken as a whole to the heavens taken as a whole. In the treatment of the individual aspects which follows, we must first discuss the position of the ecliptic and the regions of our part of the inhabited world and also the features differentiating each from the others due to the [varying] latitude at each horizon taken in order. For if the theory of these matters is treated first it will make examination of the rest easier. Secondly, we have to go through the motion of the sun and of the moon, and the phenomena accompanying these [motions]; for it would be impossible to examine the theory of the stars [and planets] thoroughly without first having a grasp of these matters. Our final task in this way of approach is the theory of the stars. Here too it would be appropriate to deal first with the sphere of the so-called 'fixed stars', and follow that by treating the five 'planets', as they are called. We shall try to provide proofs in all of these topics by using as starting-points and foundations, as it were, for our search the obvious phenomena, and those observations made by the ancients and in our own times which are reliable. We shall attach the subsequent structure of ideas to this [foundation] by means of proofs using geometrical methods.

The general preliminary discussion covers the following topics: the heaven is spherical in shape, and moves as a sphere; the earth too is sensibly spherical in shape, when taken as a whole; in position it lies in the middle of the heavens very much like its centre; in size and distance it has the ratio of a point to the sphere of the fixed stars; and it has no motion from place to place. We shall briefly discuss each of these points for the sake of reminder.

Stars Revolve in a Sphere

It is plausible to suppose that the ancients got their first notions on these topics from the following kind of observations. They saw that the sun, moon and other stars were carried from east to west along circles which were always parallel to each other, that they began to rise up from below the earth itself, as it were, gradually got up high, then kept on going round in similar fashion and getting lower, until, falling to earth, so to speak, they vanished completely, then, after remaining invisible for some time, again rose afresh and set; and [they saw] that the periods of these [motions], and also the places of rising and setting, were, on the whole, fixed and the same.

What chiefly led them to the concept of a sphere was the revolution of the ever-visible stars, which was observed to be circular, and always taking place about one centre, the same [for all]. For by necessity that point became [for them] the pole of the heavenly sphere: those stars which were closer to it revolved on smaller circles, those that were farther away described circles ever greater in proportion to their distance, until one reaches the distance of the stars which become invisible. In the case of these, too, they saw that those near the ever-visible stars remained invisible for a short time, while those farther away remained invisible for a long time, again in proportion [to their distance]. The result was that in the beginning they got to the aforementioned notion solely from such considerations; but from then on, in their subsequent investigation, they found that everything else accorded with it, since absolutely all phenomena are in contradiction to the alternative notions which have been propounded.

For if one were to suppose that the stars' motion takes place in a straight line towards infinity, as some people have thought, what device could one conceive of which would cause each of them to appear to begin their motion from the same starting-point every day? How could the stars turn back if their motion is towards infinity? Of, if they did turn back, how could this not be obvious? [On such a hypothesis], they must gradually diminish in size until they disappear, whereas, on the contrary, they are seen to be greater at the very moment of their disappearance, at which time they are gradually obstructed and cut off, as it were, by the earth's surface.

But to suppose that they are kindled as they rise out of the earth and are extinguished again as they fall to earth is a completely absurd hypothesis. For even if we were to concede that the strict order in their size and number, their intervals, positions and periods could be restored by such a random and chance process; that one whole area of the earth has a kindling nature, and another an extinguishing one, or rather that the same part [of the earth] kindles for one set of observers and extinguishes for another set; and that the same stars are already kindled or extinguished for some observers while they are not yet for others: even if, I say, we were to concede all these ridiculous consequences, what could we say about the ever-visible stars, which neither rise nor set? Those stars which are kindled and extinguished ought to rise and set for observers everywhere, while those which are not kindled and extinguished ought always to be visible for observers everywhere. What cause could we assign for the fact that this is not so? We will surely not say that stars which are kindled and extinguished for some observers never undergo this process for other observers. Yet it is utterly obvious that the same stars rise and set in certain regions [of the earth] and do neither at others.

No Variations in Stars' Movements

To sum up, if one assumes any motion whatever, except spherical, for the heavenly bodies, it necessarily follows that their distances, measured from the earth upwards, must vary, wherever and however one supposes the earth itself to be situated. Hence the sizes and mutual distances of the stars must appear to vary for the same observers during the course of each revolution, since at one time they must be at a greater distance, at another at a lesser. Yet we see that no such variation occurs. For the apparent increase in their sizes at the horizons is caused, not by a decrease in their distances, but by the exhalations of moisture surrounding the earth being interposed between the place from which we observe and the heavenly bodies, just as objects placed in water appear bigger than they are, and the lower they sink, the bigger they appear.

The following considerations also lead us to the concept of the sphericity of the heavens. No other hypothesis but this can explain how sundial constructions produce correct results; furthermore, the motion of the heavenly bodies is the most unhampered and free of all motions, and freest motion belongs among plane figures to the circle and among solid shapes to the sphere; similarly, since of different shapes having an equal boundary those with more angles are greater [in area or volume], the circle is greater than [all other] surfaces, and the sphere greater than [all other] solids; [likewise] the heavens are greater than all other bodies.

Furthermore, one can reach this kind of notion from certain physical considerations. E.g., the aether is, of all bodies, the one with constituent parts which are finest and most like each other; now bodies with parts like each other have surfaces with parts like each other; but the only surfaces with parts like each other are the circular, among planes, and the spherical, among three-dimensional surfaces. And since the aether is not plane, but three-dimensional, it follows that it is spherical in shape. Similarly, nature formed all earthly and corruptible bodies out of shapes which are round but of unlike parts, but all aethereal and divine bodies out of shapes which are of like parts and spherical. For if they were flat or shaped like a discus they would not always display a circular shape to all those observing them simultaneously from different places on earth. For this reason it is plausible that the aether surrounding them, too, being of the same nature, is spherical, and because of the likeness of its parts moves in a circular and uniform fashion.

The Earth Is Round

That the earth, too, taken as a whole, is sensibly spherical can best be grasped from the following considerations. We can see, again, that the sun, moon and other stars do not rise and set simultaneously for everyone on earth, but do so earlier for those more towards the east, later for those towards the west. For we find that the phenomena at eclipses, especially lunar eclipses, which take place at the same time [for all observers], are nevertheless not recorded as occurring at the same hour (that is at an equal distance from noon) by all observers. Rather, the hour recorded by the more easterly observers is always later than that recorded by the more westerly. We find that the differences in the hour are proportional to the distances between the places [of observation]. Hence one can reasonably conclude that the earth's surface is spherical, because its evenly curving surface (for so it is when considered as a whole) cuts off [the heavenly bodies] for each set of observers in turn in a regular fashion.

If the earth's shape were any other, this would not happen, as one can see from the following arguments. If it were concave, the stars would be seen rising first by those more towards the west; if it were plane, they would rise and set simultaneously for everyone on earth; if it were triangular or square or any other polygonal shape, by a similar argument, they would rise and set simultaneously for all those living on the same plane surface. Yet it is apparent that nothing like this takes place. Nor could it be cylindrical, with the curved surface in the east-west direction, and the flat sides towards the poles of the universe, which some might suppose more plausible. This is clear from the following: for those living on the curved surface none of the stars would be ever-visible, but either all stars would rise and set for all observers, or the same stars, for an equal [celestial] distance from each of the poles, would always be invisible for all observers. In fact, the further we travel toward the north, the more of the southern stars disappear and the more of the northern stars appear. Hence it is clear that here too the curvature of the earth cuts off [the heavenly bodies] in a regular fashion in a north-south direction, and proves the sphericity [of the earth] in all directions.

There is the further consideration that if we sail towards mountains or elevated places from and to any direction whatever, they are observed to increase gradually in size as if rising up from the sea itself in which they had previously been submerged: this is due to the curvature of the surface of the water.

At the Center of the Heavens

Once one has grasped this, if one next considers the position of the earth, one will find that the phenomena associated with it could take place only if we assume that it is in the middle of the heavens, like the centre of a sphere. For if this were not the case, the earth would have to be either

[a] not on the axis [of the universe] but equidistant from both poles, or
[b] on the axis but removed towards one of the poles, or
[c] neither on the axis nor equidistant from both poles.

Against the first of these three positions militate the following arguments. If we imagined [the earth] removed towards the zenith or the nadir of some observer, then, if he were at *sphaera recta* [at the equator], he would never experience equinox, since the horizon would always divide the heavens into two unequal parts, one above and one below the earth; if he were at *sphaera obliqua* [midway between equator and pole], either, again, equinox would never occur at all, or, [if it did occur,] it would not be at a position halfway between summer and winter solstices, since these intervals would necessarily be unequal, because the equator, which is the greatest of all parallel circles drawn about the poles of the [daily] motion, would no longer be bisected by the horizon; instead [the horizon would bisect] one of the circles parallel to the equator, either to the north or to the south of it. Yet absolutely everyone agrees that these intervals are equal everywhere on earth, since [everywhere] the increment of the longest day over the equinoctial day at the summer solstice is equal to the decrement of the shortest day from the equinoctial day at the winter solstice. But if, on the other hand, we imagined the displacement to be towards the east or west of some observer, he would find that the sizes and distances of the stars would not remain constant and unchanged at eastern and western horizons, and that the time-interval from rising to culmination would not be equal to the interval from culmination to setting. This is obviously completely in disaccord with the phenomena.

Against the second position, in which the earth is imagined to lie on the axis removed towards one of the poles, one can make the following objections. If this were so, the plane of the horizon would divide the heavens into a part above the earth and a part below the earth which are unequal and always different for different latitudes, whether one considers the relationship of the same part at two different latitudes or the two parts at the same latitude. Only at *sphaera recta* could the horizon bisect the sphere; at a *sphaera obliqua* situation such that the nearer pole were the ever-visible one, the horizon would always make the part above the earth lesser and the part below the earth greater; hence another phenomenon would be that the great circle of the ecliptic would be divided into unequal parts by the plane of the horizon. Yet it is apparent that this is by no means so. Instead, six zodiacal signs are visible above the earth at all times and places, while the remaining six are invisible; then again [at a later time] the latter are visible in their entirety above the earth, while at the same time the others are not visible. Hence it is obvious that the horizon bisects the zodiac, since the same semi-circles are cut off by it, so as to appear at one time completely above the earth, and at another [completely] below it.

And in general, if the earth were not situated exactly below the [celestial] equator, but were removed towards the north or south in the direction of one of the poles, the result would be that at the equinoxes the shadow of the gnomon [sundial] at sunrise would no longer form a straight line with its shadow at sunset in a plane parallel to the horizon, not even sensibly. Yet this is a phenomenon which is plainly observed everywhere.

Taking Account of Lunar Eclipses

It is immediately clear that the third position enumerated is likewise impossible, since the sorts of objection which we made to the first [two] will both arise in that case.

To sum up, if the earth did not lie in the middle [of the universe], the whole order of things which we observe in the increase and decrease of the length of daylight would be fundamentally upset. Furthermore, eclipses of the moon would not be restricted to situations where the moon is diametrically opposite the sun (whatever part of the heaven [the luminaries are in]), since the earth would often come between them when they were not diametrically opposite, but at intervals of less than a semi-circle.

Moreover, the earth has, to the senses, the ratio of a point to the distance of the sphere of the so-called fixed stars. A strong indication of this is the fact that the sizes and distances of the stars, at any given time, appear equal and the same from all parts of the earth everywhere, as observations of the same [celestial] objects from different latitudes are found to have not the least discrepancy from each other. One must also consider the fact that gnomons set up in any part of the earth whatever, and likewise the centres of armillary spheres [astronomical instruments], operate like the real centre of the earth; that is, the lines of sight [to heavenly bodies] and the paths of shadows caused by them agree as closely with the [mathematical] hypotheses explaining the phenomena as if they actually passed through the real centre-point of the earth.

Another clear indication that this is so is that the planes drawn through the observer's lines of sight at any point [on earth], which we call 'horizons', always bisect the whole heavenly sphere. This would not happen if the earth were of perceptible size in relation to the distance of the heavenly bodies; in that case only the plane drawn through the centre of the earth could bisect the sphere, while a plane through any point on the surface of the earth would always make the section [of the heavens] below the earth greater than the section above it.

The Earth Does Not Move

One can show by the same arguments as the preceding that the earth cannot have any motion in the aforementioned directions, or indeed ever move at all from its position at the centre. For the same phenomena would result as would if it had any position other than the central one. Hence I think it is idle to seek for causes for the motion of objects towards the centre, once it has been so clearly established from the actual phenomena that the earth occupies the middle place in the universe, and that all heavy objects are carried towards the earth. The following fact alone would most readily lead one to this notion [that all objects fall towards the centre]. In absolutely all parts of the earth, which, as we said, has been shown to be spherical and in the middle of the universe, the direction and path of the motion (I mean the proper, [natural] motion) of all bodies possessing weight is always and everywhere at right angles to the rigid plane drawn tangent to the point of impact. It is clear from this fact that, if [these falling objects] were not arrested by the surface of the earth, they would certainly reach the centre of the earth itself, since the straight line to the centre is also always at right angles to the plane tangent to the sphere at the point of intersection [of that radius] and the tangent.

Those who think it paradoxical that the earth, having such a great weight, is not supported by anything and yet does not move, seem to me to be making the mistake of judging on the basis of their own experience instead of taking into account the peculiar nature of the universe. They would not, I think, consider such a thing strange once they realised that this great bulk of the earth, when compared with the whole surrounding mass [of the universe], has the ratio of a point to it. For when one looks at it in that way, it will seem quite possible that that which is relatively smallest should be overpowered and pressed in equally from all directions to a position of equilibrium by that which is the greatest of all and of uniform nature. For there is no up and down in the universe with respect to itself, any more than one could imagine such a thing in a sphere: instead the proper and natural motion of the compound bodies in it is as follows: light and rarefied bodies drift outwards towards the circumference, but seem to move in the direction which is 'up' for each observer, since the overhead direction for all of us, which is also called 'up', points towards the surrounding surface; heavy and dense bodies, on the other hand, are carried towards the middle and the centre, but seem to fall downwards, because, again, the direction which is for all us towards our feet, called 'down', also points towards the centre of the earth. These heavy bodies, as one would expect, settle about the centre because of their mutual pressure and resistance, which is equal and uniform from all directions. Hence, too, one can see that it is plausible that the earth, since its total mass is so great compared with the bodies which fall towards it, can remain motionless under the impact of these very small weights (for they strike it from all sides), and receive, as it were, the objects falling on it. If the earth had a single motion in common with other heavy objects, it is obvious that it would be carried down faster than all of them because of its much greater size: living things and individual heavy objects would be left behind, riding on the air, and the earth itself would very soon have fallen completely out of the heavens. But such things are utterly ridiculous merely to think of.

Argument Against a Turning Earth

But certain people, [propounding] what they consider a more persuasive view, agree with the above, since they have no argument to bring against it, but think that there could be no evidence to oppose their view if, for instance, they supposed the heavens to remain motionless, and the earth to revolve from west to east about the same axis [as the heavens], making approximately one revolution each day; or if they made both heaven and earth move by any amount whatever, provided, as we said, it is about the same axis, and in such a way as to preserve the overtaking of one by the other. However, they do not realise that, although there is perhaps nothing in the celestial phenomena which would count against that hypothesis, at least from simpler considerations, nevertheless from what would occur here on earth and in the air, one can see that such a notion is quite ridiculous. Let us concede to them [for the sake of argument] that such an unnatural thing could happen as that the most rare and light of matter should either not move at all or should move in a way no different from that of matter with the opposite nature (although things in the air, which are less rare [than the heavens] so obviously move with a more rapid motion than any earthly object); [let us concede that] the densest and heaviest objects have a proper motion of the quick and uniform kind which they suppose (although, again, as all agree, earthly objects are sometimes not readily moved even by an external force). Nevertheless, they would have to admit that the revolving motion of the earth must be the most violent of all motions associated with it, seeing that it makes one revolution in such a short time; the result would be that all objects not actually standing on the earth would appear to have the same motion, opposite to that of the earth: neither clouds nor other flying or thrown objects would ever be seen moving towards the east, since the earth's motion towards the east would always outrun and overtake them, so that all other objects would seem to move in the direction of the west and the rear. But if they said that the air is carried around in the same direction and with the same speed as the earth, the compound objects in the air would none the less always seem to be left behind by the motion of both [earth and air]; or if those objects too were carried around, fused, as it were, to the air, then they would never appear to have any motion either in advance or rearwards: they would always appear still, neither wandering about nor changing position, whether they were flying or thrown objects. Yet we quite plainly see that they do undergo all these kinds of motion, in such a way that they are not even slowed down or speeded up at all by any motion of the earth.

CHAPTER 2

The Rise of Modern Astronomy

Copernicus Develops a Sun-Centered Model

By Robert Stawell Ball

In the second century Claudius Ptolemy postulated that Earth was the center of the solar system. The Ptolemaic system went unchallenged for the next fourteen hundred years. In the following selection a British astronomer, writing late in the nineteenth century, recounts the life and work of Nicolaus Copernicus (1473–1543), the man who overturned the Ptolemaic system. Having first trained to become a doctor, Copernicus at the age of twenty-seven followed his uncle's advice and entered a career in canonical law within the Catholic Church. However, Ball writes, he remained devoted to science. Beginning in 1514 Copernicus began jotting down revolutionary arguments against Ptolemy's doctrine that Earth is the still point at the center of a whirling universe. These were refined in his classic 1543 book, *De Revolutionibus Orbium Coelestium* (On the Revolutions of the Heavenly Spheres). Copernicus died shortly after its publication.

Sir Robert Stawell Ball was one of the most prolific authors of popular astronomy in the late nineteenth and early twentieth centuries. After holding positions at the Royal College of Science in Dublin and the Dunsink Observatory, in 1892 Ball became Lowndean Professor of Astronomy and Geometry at Cambridge University, as well as director of the university's observatory.

The promulgation of the accepted system of astronomy, called the Copernican system, which represents the earth as revolving on its axis and considers the sun as the centre of motion

Robert Stawell Ball, *Revolution of Astronomy by Copernicus*, 1895.

for the earth and other planets, marked the greatest of scientific revolutions.

Copernicus, whose name, thus Latinized, was Koppernigk or Kopernik, was born at Thorn, Prussia, February 19, 1473, and died at Frauenburg, Prussia, May 24, 1543. The founder of modern astronomy was probably of German descent: according to some authorities his father was a Germanized Slav, his mother a German; and the honor of producing him is claimed by both Germany and Poland. . . .

Copernicus, the astronomer, whose discoveries make him the great predecessor of [Johannes] Kepler and [Isaac] Newton, did not come from a noble family, as certain other early astronomers have done, for his father was a tradesman. Chroniclers are, however, careful to tell us that one of his uncles was a bishop. We are not acquainted with any of those details of his childhood or youth which are often of such interest in other cases where men have risen to exalted fame. It would appear that the young Nicolaus, for such was his Christian name, received his education at home until such time as he was deemed sufficiently advanced to be sent to the University of Cracow. The education that he there obtained must have been in those days of very primitive description, but Copernicus seems to have availed himself of it to the utmost. He devoted himself more particularly to the study of medicine, with the view of adopting its practice as the profession of his life. The tendencies of the future astronomer were, however, revealed in the fact that he worked hard at mathematics, and for him, as for one of his illustrious successors, Galileo, the practice of the art of painting had a very great interest, and in it he obtained some measure of success.

Devoted to Science and Church

By the time he was twenty-seven years old, it would seem that Copernicus had given up the notion of becoming a medical practitioner, and had resolved to devote himself to science. He was engaged in teaching mathematics, and appears to have acquired some reputation. His growing fame attracted the notice of his uncle the Bishop, at whose suggestion Copernicus took holy orders, and he was presently appointed to a canonry in the Cathedral of Frauenburg, near the mouth of the Vistula.

To Frauenburg, accordingly, this man of varied gifts retired. Possessing somewhat of the ascetic spirit, he resolved to devote his life to work of the most serious description. He eschewed all ordinary society, restricting his intimacies to very grave and learned companions, and refusing to engage in conversation of any useless kind. It would seem as if his gifts for painting were condemned as frivolous; at all events, we do not learn that he continued to practise them. In addition to the discharge of his theological duties, his life was occupied partly in ministering medically to the wants of the poor, and partly with his researches in astronomy and mathematics. His equipment in the matter of instruments for the study of the heavens seems to have been of a very meagre description. He arranged apertures in the walls of his house at Allenstein, so that he could observe in some fashion the passage of the stars across the meridian. That he possessed some talent for practical mechanics is proved by his construction of a contrivance for raising water from a stream, for the use of the inhabitants of Frauenburg. Relics of this machine are still to be seen.

The intellectual slumber of the Middle Ages was destined to be awakened by the revolutionary doctrines of Copernicus. It may be noted, as an interesting circumstance, that the time at which he discovered the scheme of the solar system coincided with a remarkable epoch in the world's history. The great astronomer had just reached manhood at the time when Columbus discovered the New World.

Ptolemy's Doctrines

Before the publication of the researches of Copernicus, the orthodox scientific creed averred that the earth was stationary, and that the apparent movements of the heavenly bodies were real movements. [Greek astronomer Claudius] Ptolemy had laid down this doctrine fourteen hundred years before. In his theory this huge error was associated with so much important truth, and the whole presented such a coherent scheme for the explanation of the heavenly movements, that the Ptolemaic theory was not seriously questioned until the great work of Copernicus appeared. No doubt others before Copernicus had from time to time in some vague fashion surmised, with more or less plausibility, that the sun, and not the earth, was the centre about which the system really revolved. It is, however, one thing to state a scientific fact; it is quite another thing to be in possession of the train of reasoning, founded on observation or experiment, by which that fact may be established. Pythagoras, it appears, had indeed told his disciples that it was the sun, and not the earth, which was the centre of movement, but it does not seem at all certain that Pythagoras had any grounds which science could recognize for the belief which is attributed to him. So far as information is available to us, it would seem that Pythagoras associated his scheme of things celestial with a number of preposterous notions in natural philosophy. He may certainly have made a correct statement as to which was the most important body in the solar system, but he certainly did not provide any rational demonstration of the fact. Copernicus, by a strict train of reasoning, convinced those who would listen to him that the sun was the centre of the system. It is useful for us to consider the arguments which he urged and by which he effected that intellectual revolution which is always connected with his name.

Earth's Rotation

The first of the great discoveries which Copernicus made relates to the rotation of the earth on its axis. That general diurnal movement, by which the stars and all other celestial bodies appear to be carried completely round the heavens once every twenty-four hours, had been accounted for by Ptolemy on the supposition that the apparent movements were the real movements. Ptolemy himself felt the extraordinary difficulty involved in the supposition that so stupendous a fabric as the celestial sphere should spin in the way supposed. Such movements required that many of the stars should travel with almost inconceivable velocity. Copernicus also saw that the daily rising and setting of the heavenly bodies could be accounted for either by the supposition that the celestial sphere moved round and that the earth remained at rest, or by the supposition that the celestial sphere was at rest while the earth turned round in the opposite direction. He weighed the arguments on both sides as Ptolemy had done, and as the result of his deliberation Copernicus came to an opposite conclusion from Ptolemy. To Copernicus it appeared that the difficulties attending the supposition that the celestial sphere revolved were vastly greater than those which appeared so weighty to Ptolemy as to force him to deny the earth's rotation.

Copernicus shows clearly how the observed phenomena could be accounted for just as completely by a rotation of the earth as by a rotation of the heavens. He alludes to the fact that, to those on board a vessel which is moving through smooth water, the vessel itself appears to be at rest, while the objects on shore appear to be moving past. If, therefore, the earth were rotating uniformly, we dwellers upon the earth, oblivious of our own movement, would wrongly attribute to the stars the displacement which was actually the consequence of our own motion.

Copernicus saw the futility of the arguments by which Ptolemy had endeavored to demonstrate that a revolution of the earth was impossible. It was plain to him that there was nothing whatever to warrant refusal to believe in the rotation of the earth. In his clear-sightedness on this matter we have specially to admire the sagacity of Copernicus as a natural philosopher. It had been urged that, if the earth moved round, its motion would not be imparted to the air, and that therefore the earth would be uninhabitable by the terrific winds which would be the result of our being carried through the air. Copernicus convinced himself that this deduction was preposterous. He proved that the air must accompany the earth, just as one's coat remains round him, notwithstanding the fact that he is walking down the street. In this way he was able to show that all *a priori* objections to the earth's movements were absurd, and therefore he was able to compare together the plausibilities of the two rival schemes for explaining the diurnal movement.

Disputing the Celestial Sphere

Once the issue had been placed in this form, the result could not be long in doubt. Here is the question: Which is it more likely— that the earth, like a grain of sand at the centre of a mighty globe, should turn round once in twenty-four hours, or that the whole of that vast globe should complete a rotation in the opposite direction in the same time? Obviously, the former is far the more simple supposition. But the case is really much stronger than this. Ptolemy had supposed that all the stars were attached to the surface of a sphere. He had no ground whatever for this supposition, except that otherwise it would have been wellnigh impossible to devise a scheme by which the rotation of the heavens around a fixed earth could have been arranged. Copernicus, however, with the just instinct of a philosopher, considered that the celestial sphere, however convenient, from a geometrical point of view, as a means of representing apparent phenomena, could not actually have a material existence. In the first place, the existence of a material celestial sphere would require that all the myriad stars should be at exactly the same distances from the earth. Of course, no one will say that this or any other arbitrary disposition of the stars is actually impossible; but as there was no conceivable physical reason why the distances of all the stars from the earth should be identical, it seemed in the very highest degree improbable that the stars should be so placed.

Doubtless, also, Copernicus felt a considerable difficulty as to the nature of the materials from which Ptolemy's wonderful sphere was to be constructed. Nor could a philosopher of his pen-

In the sixteenth century Copernicus proposed that the sun, not the earth, was the center of the universe.

etration have failed to observe that, unless that sphere were infinitely large, there must have been space outside it, a consideration which would open up other difficult questions. Whether infinite or not, it was obvious that the celestial sphere must have a diameter at least many thousands of times as great as that of the earth. From these considerations Copernicus deduced the important fact that the stars and other important celestial bodies must all be vast objects. He was thus enabled to put the question in such a form that it would hardly receive any answer but the correct one: Which is it more rational to suppose, that the earth should turn round on its axis once in twenty-four hours, or that thousands of mighty stars should circle round the earth in the same time, many of them having to describe circles many thousands of times greater in circumference than the circuit of the earth at the equator? The obvious answer pressed upon Copernicus with so much force that he was compelled to reject Ptolemy's theory of the stationary earth, and to attribute the diurnal rotation of the heavens to the revolution of the earth on its axis.

Once this tremendous step had been taken, the great difficulties which beset the monstrous conception of the celestial sphere vanished, for the stars need no longer be regarded as situated at equal distances from the earth. Copernicus saw that they might lie at the most varied degrees of remoteness, some being hundreds or thousands of times farther away than others. The complicated structure of the celestial sphere as a material object disappeared altogether; it remained only as a geometrical conception, whereon we find it convenient to indicate the places of the stars. Once the Copernican doctrine had been fully set forth, it was impossible for anyone, who had both the inclination and the capacity to understand it, to withhold acceptance of its truth. The doctrine of a stationary earth had gone forever.

Tackling Planetary Movements

Copernicus having established a theory of the celestial movements which deliberately set aside the stability of the earth, it seemed natural that he should inquire whether the doctrine of a moving earth might not remove the difficulties presented in other celestial phenomena. It had been universally admitted that the earth lay unsupported in space. Copernicus had further shown that it possessed a movement of rotation. Its want of stability being thus recognized, it seemed reasonable to suppose that the earth might also have some other kinds of movements as well. In this, Copernicus essayed to solve a problem far more difficult than that which hitherto occupied his attention. It was a comparatively easy task to show how the diurnal rising and setting could be accounted for by the rotation of the earth. It was a much more difficult undertaking to demonstrate that the planetary movements, which Ptolemy had represented with so much success, could be completely explained by the supposition that each of these planets revolved uniformly round the sun, and that the earth was also a planet, accomplishing a complete circuit of the sun once in the course of a year.

It would be impossible, in a sketch like the present, to enter into any detail as to the geometrical propositions on which this beautiful investigation of Copernicus depended. We can only mention a few of the leading principles. It may be laid down in general that, if an observer is in movement, he will, if unconscious of the fact, attribute to the fixed objects around him a movement equal and opposite to that which he actually possesses. A passenger on a canal-boat sees the objects on the banks apparently moving backward with a speed equal to that by which he himself is advancing forward. By an application of this principle, we can account for all the phenomena of the movements of the planets, which Ptolemy had so ingeniously represented by his circles. Let us take, for instance, the most characteristic feature in the irregularities of the outer planets. Mars, though generally advancing from west to east among the stars, occasionally pauses, retraces his steps for a while, again pauses, and then resumes his ordinary onward progress. Copernicus showed clearly how this effect was produced by the real motion of the earth, combined with the real motion of Mars. When the earth comes directly between Mars and the sun, the retrograde movement of Mars is at its highest. Mars and the earth are then advancing in the same direction. We, on the earth, however, being unconscious of our own motion, attribute, by the principle I have already explained, an equal and opposite motion to Mars. The visible effect upon the planet is that Mars has two movements, a real onward movement in one direction, and an apparent movement in the opposite direction. If it so happened that the earth was moving with the same speed as Mars, then the apparent movement would exactly neutralize the real movement, and Mars would seem to be at rest relatively to the surrounding stars. Under the actual circumstances considered, however, the earth is moving faster than Mars, and the consequence is that the apparent movement of the planet backward exceeds the real movement forward, the net result being an apparent retrograde movement.

Heliocentric System Complete

With consummate skill, Copernicus showed how the applications of the same principles could account for the characteristic movements of the planets. His reasoning in due time bore down all opposition. The supreme importance of the earth in the system vanished. It had now merely to take rank as one of the planets.

The same great astronomer now, for the first time, rendered something like a rational account of the changes of the seasons. Nor did certain of the more obscure astronomical phenomena escape his attention.

He delayed publishing his wonderful discoveries to the world until he was quite an old man. He had a well-founded apprehension of the storm of opposition which they would arouse. However, he yielded at last to the entreaties of his friends, and his book was sent to the press. But ere it made its appearance to the world, Copernicus was seized with mortal illness. A copy of the book was brought to him on May 23, 1543. We are told that he was able to see it and to touch it, but no more; and he died a few hours afterward.

The Advent of the Telescope

By Philip S. Harrington

Historians disagree about the invention of the telescope. Many credit Jan Lippershey of the Netherlands with having invented the two-lens telescope in 1608; some consider his contemporary and countryman, Jacob Metius, to be the inventor; still others believe it may have been invented considerably earlier.

In the following selection Philip S. Harrington offers the view that no one knows for certain who invented the device. What is certain, however, is that the Italian scientist Galileo Galilei made his own refracting telescope in 1609 and made it popular by reporting on his telescopic observations of the planets. A few years later the German astronomer Johannes Kepler made an improvement on Galileo's design, replacing the concave eyepiece lens with a double convex one. However, even Kepler's model caused severe distortions because the large lens in the front of the telescope focused different wavelengths at different points, a problem called chromatic distortion. The shape of the lens—being a section of a sphere—also caused distortions (spherical aberrations) around the edges. Over the next few centuries these problems were greatly reduced by the invention of lenses with different kinds of glass sandwiched together.

In the meantime, another kind of telescope was invented. In 1663 Scottish mathematician James Gregory designed the first reflecting telescope, which employs a concave mirror to collect light and focus it on a second mirror, which reflects it into an eyepiece for viewing. After many modifications, the reflecting telescope became the standard for large observatories, although many amateur astronomers continue to use refracting telescopes or a combination of the two. Author Philip S. Harrington was trained in mechanical engineering and science education. He has published several books on astronomical viewing.

Philip S. Harrington, *Star Ware: The Amateur Astronomer's Ultimate Guide to Choosing, Buying, and Using Telescopes and Accessories.* New York: John Wiley & Sons, 2002. Copyright © 2002 by Philip S. Harrington. All rights reserved. Reproduced by permission of John Wiley & Sons, Inc.

Since its invention, the telescope has captured the curiosity and commanded the respect of princes and paupers, scientists and laypersons. Peering through a telescope renews the sense of wonder we all had as children. In short, it is a tool that sparks the imagination in us all.

Who is responsible for this marvelous creation? Ask this question of most people and they probably will answer, "Galileo." [Seventeenth-century Italian scientist] Galileo Galilei did, in fact, usher in the age of telescopic astronomy when he first turned his telescope . . . toward the night sky [in 1609]. With it, he became the first person in human history to witness craters on the Moon, the phases of Venus, four of the moons orbiting Jupiter, and also many other hitherto unknown heavenly sights. Though he was ridiculed by his contemporaries and persecuted for heresy, Galileo's observations changed humankind's view of the universe as no single individual ever had before or has since. But he did not make the first telescope.

So who did? The truth is that no one knows for certain just who came up with the idea, or even when. Many knowledgeable historians tell us that it was Jan Lippershey, a spectacle maker from Middelburg, Holland. Records indicate that in 1608 he first held two lenses in line and noticed that they seemed to bring distant scenes closer. Subsequently, Lippershey sold many of his telescopes to his government, which recognized the military importance of such a tool. In fact, many of his instruments were sold in pairs, thus creating the first field glasses.

Ancient Glass

Other evidence may imply a much earlier origin for the telescope. Archaeologists have unearthed glass in Egypt that dates to about 3500 B.C., while primitive lenses have been found in Turkey and Crete that are thought to be 4,000 years old! In the third century B.C., Euclid wrote about the reflection and refraction of light. Four hundred years later, the Roman writer Seneca referred to the magnifying power of a glass sphere filled with water.

Although it is unknown if any of these independent works led to the creation of a telescope, the English scientist Roger Bacon wrote of an amazing observation made in the thirteenth century: ". . . Thus from an incredible distance we may read the smallest letters . . . the Sun, Moon and stars may be made to descend hither

in appearance. . . ." Might he have been referring to the view
through a telescope? We may never know.

Refracting Telescopes

Though its inventor may be lost to history, this early kind of telescope is called a *Galilean* or *simple* refractor. The Galilean refractor consists of two lenses: a convex (curved outward) lens held in
front of a concave (curved inward) lens a certain distance away.
As you [might] know, the telescope's front lens is called the objective, while the other is referred to as the eyepiece, or *ocular.*
The Galilean refractor placed the concave eyepiece *before* the objective's prime focus; this produced an upright, extremely narrow
field of view, like today's inexpensive opera glasses.

Not long after Galileo made his first telescope, [German as-
tronomer] Johannes Kepler improved on the idea by simply swap-
ping the concave eyepiece for a double convex lens, placing it behind the prime focus. The *Keplerian refractor* proved to be far
superior to Galileo's instrument. The modern refracting telescope
continues to be based on Kepler's design. The fact that the view
is upside down is of little consequence to astronomers because
there is no up and down in space; for terrestrial viewing, extra
lenses may be added to flip the image a second time, reinverting
the scene.

Problems of Distortion

Unfortunately, both the Galilean and the Keplerian designs have
several optical deficiencies. Chief among these is *chromatic aberration.* As you may know, when we look at any white-light source,
we are not actually looking at a single wavelength of light but
rather a collection of wavelengths mixed together. To prove this
for yourself, shine sunlight through a prism. The light going in is
refracted within the prism, exiting not as a unit but instead broken
up, forming a rainbowlike spectrum. Each color of the spectrum
has its own unique wavelength.

If you use a lens instead of a prism, each color will focus at a
slightly different point. The net result is a zone of focus, rather
than a point. Through such a telescope, everything appears blurry
and surrounded by halos of color. This effect is called chromatic
aberration.

Another problem of simple refractors is *spherical aberration.* In this instance, the curvature of the objective lens causes the rays of light entering around its edges to focus at a slightly different place than those striking the center. Once again, the light focuses within a range rather than at a single point, making the telescope incapable of producing a clear, razor-sharp image.

Modifying the inner and outer curves of the lens proved somewhat helpful. Experiments showed that both defects could be reduced (but not totally eliminated) by increasing the focal length—that is, decreasing the curvature—of the objective lens. And so, in an effort to improve image quality, the refractor became longer . . . and longer . . . and even longer! The longest refractor on record was constructed by Johannes Hevelius in Denmark during the latter part of the seventeenth century; it measured about one hundred and fifty feet from objective to eyepiece and required a complex sling system suspended high above the ground on a wooden mast to hold it in place! Can you imagine the effort it must have taken to swing around such a monster just to look at the Moon or a bright planet? Surely, there had to be a better way.

Double Lens Solution

In an effort to combat these imperfections, Chester Hall developed a two-element *achromatic lens* in 1733. Hall learned that by using two matching lenses made of different types of glass, aberrations could be greatly reduced. In an achromatic lens, the outer element is usually made of crown glass, while the inner element is typically flint glass. Crown glass has a lower dispersion effect and therefore bends light rays less than flint glass, which has a higher dispersion. The convergence of light passing through the crown-glass lens is compensated by its divergence through the flint-glass lens, resulting in greatly dampened aberrations. Ironically, though Hall made several telescopes using this arrangement, the idea of an achromatic objective did not catch on for another quarter century.

In 1758, [British lens maker] John Dollond reacquainted the scientific community with Hall's idea when he was granted a patent for a two-element aberration-suppressing lens. Though quality glass was hard to come by for both of these pioneers, it appears that Dollond was more successful at producing a high-quality instrument. Perhaps that is why history records John Dollond, rather than Chester Hall, as the father of the modern refractor.

Regardless of who first devised it, this new and improved design has come to be called the *achromatic refractor*, with the compound objective simply labeled an *achromat*. Though the methodology for improving the refractor was now known, the problem of getting high-quality glass (especially flint glass) persisted. In 1780, Pierre Louis Guinard, a Swiss bell maker, began experimenting with various casting techniques in an attempt to improve the glass-making process. It took him close to 20 years, but Guinard's efforts ultimately paid off, for he learned the secret of producing flawless optical disks as big as roughly 6 inches in diameter.

Later, Guinard was to team up with Joseph von Fraunhofer, inventor of the spectroscope. While studying under Guinard's guidance, Fraunhofer experimented by slightly modifying the lens curves suggested by Dollond, which resulted in the highest-quality objective yet created. In Fraunhofer's design, the front surface is strongly convex. The two central surfaces differ slightly from each other, requiring a narrow air space between the elements, while the innermost surface is almost perfectly flat. These innovations bring two wavelengths of light across the lens's full diameter to a common focus, thereby greatly reducing chromatic and spherical aberration. . . .

Reflecting Telescopes

But there is more than one way to skin a cat. The second general type of telescope utilizes a large mirror, rather than a lens, to focus light to a point—not just any mirror, mind you, but a mirror with a precisely figured surface. To understand how a mirror-based telescope works, we must first reflect on how mirrors work (sorry about that). Take a look at a mirror in your home. Chances are it is flat. . . . Light that is cast onto the mirror's polished surface in parallel rays is reflected back in parallel rays. If the mirror is convex, the light diverges after it strikes the surface. But if the mirror is concave, then the rays converge toward a common point, or focus. (It should be pointed out here that household mirrors are *second-surface* mirrors; that is, their reflective coating is applied onto the back surface. Reflecting telescopes use *front-surface* mirrors, coated on the front.)

The first reflecting telescope was designed by [Scottish mathematician] James Gregory in 1663. His system centered around a concave mirror (called the *primary mirror*). The primary mirror reflected light to a smaller concave *secondary mirror*, which, in turn, bounced the light back through a central hole in the primary and out to the eyepiece. The *Gregorian reflector* had the benefit of yielding an upright image, but its optical curves proved difficult for Gregory and his contemporaries to fabricate.

Newton's Reflector

A second design was later conceived by Sir Isaac Newton in 1672. Like Gregory, Newton realized that a concave mirror would reflect and focus light back along the optical axis to a point called the prime focus. Here an observer could view a magnified image through an eyepiece. Quickly realizing that his head got in the way, Newton inserted a flat mirror at a 45° angle some distance in front of the primary. The secondary, or *diagonal*, mirror acted to bounce the light 90° out through a hole in the side of the telescope's tube. This arrangement has since become known as the *Newtonian reflector....*

The French sculptor Sieur Cassegrain also announced in 1672 a third variation of the reflecting telescope. The Cassegrainian reflector (yes, the telescope is correctly called a *Cassegrainian*, but since most other sources refer to it as a *Cassegrain*, I will from this point on, as well) is outwardly reminiscent of Gregory's original design. The biggest difference between a *Cassegrain reflector* and a Gregorian reflector is the curve of the secondary mirror's surface. The Gregorian uses a concave secondary mirror positioned outside the main focus, while Cassegrain uses a convex secondary mirror inside the main focus. The biggest plus to the Cassegrain is its compact design, which combines a large aperture inside a short tube. Optical problems include lower image contrast than a Newtonian, as well as strong curvature of field and coma, causing stars along the edges of the field to blur when those in the center are focused.

Both Newton and Cassegrain received acclaim for their independent inventions, but neither telescope saw further development for many years. One of the greatest difficulties to overcome was the lack of information on suitable materials for mirrors. Newton, for instance, made his mirrors out of bell metal whitened with arsenic. Others chose speculum metal, an amalgam consisting of copper, tin, and arsenic....

Sir William Herschel, a musician who became interested in astronomy when he was given a telescope in 1722, ground some of the finest mirrors of his day. As his interest in telescopes grew, Herschel continued to refine the reflector by devising his own system. The *Herschelian* design called for the primary mirror to be tilted slightly, thereby casting the reflection toward the front rim of the oversized tube, where the eyepiece would be mounted. The biggest advantage to this arrangement is that with no secondary mirror to block the incoming light, the telescope's aperture is unobstructed by a second mirror; disadvantages included image distortion due to the tilted optics and heat from the observer's head. Herschel's largest telescope was completed in 1789. The metal speculum around which it was based measured 48 inches across and had a focal length of 40 feet. Records indicate that it weighed something in excess of one ton.

The Planets Travel Around the Sun

By Galileo Galilei

Galileo Galilei (1564–1642) discovered the moons of Jupiter and the rings of Saturn, established new laws of motion, and championed the heliocentric theory of Copernicus. In his own time in Italy, Galileo was recognized as an unsurpassed mathematician, physicist, and astronomer. The powerful Medici family backed him, and he rose to become the head of mathematics and physics at the University of Pisa. He put his career and his life at risk, however, in publishing the book from which the following excerpt comes. The book consists of an extended discourse among three characters: Salviati, an exponent of science (and stand-in for Galileo himself); Sagredo, a wise but neutral character; and Simplicio, who represents the prevailing (and in Galileo's view, foolish) Aristotelian philosophy of the day. The selection starts with Simplicio claiming that he has a hundred proofs for his view of a spherical (and Earth-centered) universe. Salviati sets out to prove to him that the Copernican system is right. He notes that the apparent size of the planets changes over time, which could not happen if they were orbiting Earth in circles. He then lays out a sun-centered system, with only the moon orbiting Earth. He goes on to argue that Earth, rather than the whole universe of stars, revolves every twenty-four hours.

Within months of its publication the *Dialogue Concerning the Two Chief World Systems* was called heretical by the Vatican and banned. In 1633 Galileo faced trial before the Inquisition. Under threat of torture, Galileo recanted his claim that Earth moves around the sun. Spared capital punishment, he was nevertheless sentenced to house arrest, which continued for the remainder of his life. In 1992 Pope John Paul II acknowledged that the Church had been wrong and lifted the edict against Galileo.

Galileo Galilei, *Dialogue Concerning the Two Chief World Systems—Ptolemaic & Copernican*, translated by Stillman Drake. Berkeley: University of California Press, 1967.

SIMPLICIO: Aristotle gives a hundred proofs that the universe is finite, bounded, and spherical.

SALVIATI: Which are later all reduced to one, and that one to none at all. For if I deny him his assumption that the universe is movable all his proofs fall to the ground, since he proves it to be finite and bounded only if the universe is movable. But in order not to multiply our disputes, I shall concede to you for the time being that the universe is finite, spherical, and has a center. And since such a shape and center are deduced from mobility, it will be the more reasonable for us to proceed from this same circular motion of world bodies to a detailed investigation of the proper position of the center. Even Aristotle himself reasoned about and decided this in the same way, making that point the center of the universe about which all the celestial spheres revolve, and at which he believed the terrestrial globe to be situated. Now tell me, Simplicio: if Aristotle had found himself forced by the most palpable experiences to rearrange in part this order and disposition of the universe, and to confess himself to have been mistaken about one of these two propositions—that is, mistaken either about putting the earth in the center, or about saying that the celestial spheres move around such a center—which of these admissions do you think that he would choose?

SIMPLICIO: I think that if that should happen, the Peripatetics [students of Aristotle] . . .

SALVIATI: I am not asking the Peripatetics; I am asking Aristotle himself. As for the former, I know very well what they would reply. They, as most reverent and most humble slaves of Aristotle, would deny all the experiences and observations in the world, and would even refuse to look at them in order not to have to admit them, and they would say that the universe remains just as Aristotle has written; not as nature would have it. For take away the prop of his authority, and with what would you have them appear in the field? So now tell me what you think Aristotle himself would do.

Taking Issue with Aristotle

SIMPLICIO: Really, I cannot make up my mind which of these two difficulties he would have regarded as the lesser.

SALVIATI: Please, do not apply this term "difficulty" to something that may necessarily be so; wishing to put the earth in the center of the celestial revolutions was a "difficulty." But since you do not know to which side he would have leaned, and considering him as I do a man of brilliant intellect, let us set about examining which of the two choices is the more reasonable, and let us take that as the one which Aristotle would have embraced. So, resuming our reasoning once more from the beginning, let us assume out of respect for Aristotle that the universe (of the magnitude of which we have no sensible information beyond the fixed stars), like anything that is spherical in shape and moves circularly, has necessarily a center for its shape and for its motion. Being certain, moreover, that within the stellar sphere there are many orbs one inside another, with their stars which also move circularly, our question is this: Which is it more reasonable to believe and to say; that these included orbs move around the same center as the universe does, or around some other one which is removed from that? Now you, Simplicio, say what you think about this matter.

SIMPLICIO: If we could stop with this one assumption and were sure of not running into something else that would disturb us, I should think it would be much more reasonable to say that the container and the things it contained all moved around one common center rather than different ones.

The Sun at the Center

SALVIATI: Now if it is true that the center of the universe is that point around which all the orbs and world bodies (that is, the planets) move, it is quite certain that not the earth, but the sun, is to be found at the center of the universe. Hence, as for this first general conception, the central place is the sun's, and the earth is to be found as far away from the center as it is from the sun.

SIMPLICIO: How do you deduce that it is not the earth, but the sun, which is at the center of the revolutions of the planets?

SALVIATI: This is deduced from most obvious and therefore most powerfully convincing observations. The most palpable of these, which excludes the earth from the center and places the sun there, is that we find all the planets closer to the earth at one time and farther from it at another. The differences are so great that Venus, for example, is six times as distant from us at its farthest as at its closest, and Mars soars nearly eight times as high in the one state as in the other. You may thus see whether Aristotle was not some trifle deceived in believing that they were always equally distant from us.

SIMPLICIO: But what are the signs that they move around the sun?

SALVIATI: This is reasoned out from finding the three outer planets—Mars, Jupiter, and Saturn—always quite close to the earth when they are in opposition to the sun, and very distant when they are in conjunction with it. This approach and recession is of such moment that Mars when close looks sixty times as large as when it is most distant. Next, it is certain that Venus and Mercury must revolve around the sun, because of their never moving far away from it, and because of their being seen now beyond it and now on this side of it, as Venus's changes of shape conclusively prove. As to the moon, it is true that this can never separate from the earth in any way, for reasons that will be set forth more specifically as we proceed.

SAGREDO: I have hopes of hearing still more remarkable things arising from this annual motion of the earth than were those which depended upon its diurnal rotation.

SALVIATI: You will not be disappointed, for as to the action of the diurnal motion upon celestial bodies, it was not and could not be anything different from what would appear if the universe were to rush speedily in the opposite direction. But this annual motion, mixing with the individual motions of all the planets, produces a great many oddities which in the past have baffled all the greatest men in the world.

The Moon Circles the Earth

Now returning to these first general conceptions, I repeat that the center of the celestial rotation for the five planets, Saturn, Jupiter, Mars, Venus, and Mercury, is the sun; this will hold for the earth too, if we are successful in placing that in the heavens. Then as to the moon, it has a circular motion around the earth, from which as I have already said it cannot be separated; but this does not keep it from going around the sun along with the earth in its annual movement.

SIMPLICIO: I am not yet convinced of this arrangement at all. Perhaps I should understand it better from the drawing of a diagram, which might make it easier to discuss.

SALVIATI: That shall be done. But for your greater satisfaction and your astonishment, too, I want you to draw it yourself. You will see that however firmly you may believe yourself not to understand it, you do so perfectly, and just by answering my questions you will describe it exactly. So take a sheet of paper and the compasses; let this page be the enormous expanse of the universe, in which you have to distribute and arrange its parts as reason shall direct you. And first, since you are sure without my telling you that the earth is located in this universe, mark some point at your pleasure where you intend this to be located, and designate it by means of some letter.

SIMPLICIO: Let this be the place of the terrestrial globe, marked A.

SALVIATI: Very well. I know in the second place that you are aware that this earth is not inside the body of the sun, nor even contiguous to it, but is distant from it by a certain space. Therefore assign to the sun some other place of your choosing, as far from the earth as you like, and designate that also.

SIMPLICIO: Here I have done it; let this be the sun's position, marked O.

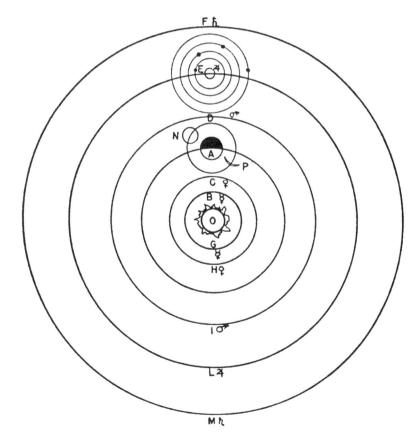

SALVIATI: These two established, I want you to think about placing Venus in such a way that its position and movement can conform to what sensible experience shows us about it. Hence you must call to mind, either from past discussions or from your own observations, what you know happens with this star. Then assign it whatever place seems suitable for it to you.

Locating the Planets

SIMPLICIO: I shall assume that those appearances are correct which you have related and which I have read also in the booklet of theses; that is, that this star never recedes from the sun beyond a certain definite interval of forty degrees or so; hence it not only never reaches opposition to the sun, but not even quadrature, nor so much as a sextile aspect. Moreover, I shall assume that it displays itself to us about forty times as large at one time than at another; greater when, being retrograde, it is approaching evening conjunction with the sun, and very small when it is moving forward toward morning conjunction, and furthermore that when it appears very large, it reveals itself in a horned shape, and when it looks very small it appears perfectly round.

These appearances being correct, I say, I do not see how to escape affirming that this star revolves in a circle around the sun, in such a way that this circle cannot possibly be said to embrace and contain within itself the earth, nor to be beneath the sun (that is, between the sun and the earth), nor yet beyond the sun. Such a circle cannot embrace the earth because then Venus would sometimes be in opposition to the sun; it cannot be beneath the sun, for then Venus would appear sickle-shaped at both conjunctions; and it cannot be beyond the sun, since then it would always look round and never horned. Therefore for its lodging I shall draw the circle CH around the sun, without having this include the earth.

SALVIATI: Venus provided for, it is fitting to consider Mercury, which, as you know, keeping itself always around the sun, recedes there from much less than Venus. Therefore consider what place you should assign to it.

SIMPLICIO: There is no doubt that, imitating Venus as it does, the most appropriate place for it will be a smaller circle, within this one of Venus and also described about the sun. A reason for this, and especially for its proximity to the sun, is the vividness of Mercury's splendor surpassing that of Venus and all the other planets.

Hence on this basis we may draw its circle here and mark it with the letters BG.

Mars and Beyond

SALVIATI: Next, where shall we put Mars?

SIMPLICIO: Mars, since it does come into opposition with the sun, must embrace the earth with its circle. And I see that it must also embrace the sun; for, coming into conjunction with the sun, if it did not pass beyond it but fell short of it, it would appear horned as Venus and the moon do. But it always looks round; therefore its circle must include the sun as well as the earth. And since I remember your having said that when it is in opposition to the sun it looks sixty times as large as when in conjunction, it seems to me that this phenomenon will be well provided for by a circle around the sun embracing the earth, which I draw here and mark DI. When Mars is at the point D, it is very near the earth and in opposition to the sun, but when it is at the point I, it is in conjunction with the sun and very distant from the earth.

And since the same appearances are observed with regard to Jupiter and Saturn (although with less variation in Jupiter than in Mars, and with still less in Saturn than in Jupiter), it seems clear to me that we can also accommodate these two planets very neatly with two circles, still around the sun. This first one, for Jupiter, I mark EL; the other, higher, for Saturn, is called FM.

SALVIATI: So far you have comported yourself uncommonly well. And since, as you see, the approach and recession of the three outer planets is measured by double the distance between the earth and the sun, this makes a greater variation in Mars than in Jupiter because the circle DI of Mars is smaller than the circle EL of Jupiter. Similarly, EL here is smaller than the circle FM of Saturn, so the variation is still less in Saturn than in Jupiter, and this corresponds exactly to the appearances. It now remains for you to think about a place for the moon.

The Moon's Orbit

SIMPLICIO: Following the same method (which seems to me very convincing), since we see the moon come into conjunction and opposition with the sun, it must be admitted that its circle embraces the earth. But it must not embrace the sun also, or else when it was in conjunction it would not look horned but always round and full of light. Besides, it would never cause an eclipse of the sun for us, as it frequently does, by getting in between us and the sun. Thus one must assign to it a circle around the earth, which shall be this one, NP, in such a way that when at P it appears to us here on the earth A as in conjunction with the sun, which sometimes it will eclipse in this position. Placed at N, it is seen in opposition to the sun, and in that position it may fall under the earth's shadow and be eclipsed.

SALVIATI: Now what shall we do, Simplicio, with the fixed stars? Do we want to sprinkle them through the immense abyss of the universe, at various distances from any predetermined point, or place them on a spherical surface extending around a center of their own so that each of them will be the same distance from that center?

SIMPLICIO: I had rather take a middle course, and assign to them an orb described around a definite center and included between two spherical surfaces—a very distant concave one, and another closer and convex, between which are placed at various altitudes the innumerable host of stars. This might be called the universal sphere, containing within it the spheres of the planets which we have already designated.

SALVIATI: Well, Simplicio, what we have been doing all this while is arranging the world bodies according to the Copernican distribution, and this has now been done by your own hand. Moreover, you have assigned their proper movements to them all except the sun, the earth, and the stellar sphere. To Mercury and Venus you have attributed a circular motion around the sun without embracing the earth. Around the same sun you have caused the three outer planets, Mars, Jupiter, and Saturn, to move, embracing the earth within their circles. Next, the moon cannot move in any way except around the earth and without embracing the sun. And in all these movements you likewise agree with Copernicus himself. It now remains to apportion three things among the sun, the earth, and the stellar sphere: the state of rest, which appears to belong to the earth; the annual motion through the zodiac, which appears to belong to the sun; and the diurnal movement, which appears to belong to the stellar sphere, with all the rest of the universe sharing in it except the earth. And since it is true that all the planetary orbs (I mean Mercury, Venus, Mars, Jupiter, and Saturn) move around the sun as a center, it seems most reasonable for the state of rest to belong to the sun rather than to the earth—just as it does for the center of any movable sphere to remain fixed, rather than some other point of it remote from the center.

The Earth Spins

Next as to the earth, which is placed in the midst of moving objects—I mean between Venus and Mars, one of which makes its revolution in nine months and the other in two years—a motion requiring one year may be attributed to it much more elegantly than a state of rest, leaving the latter for the sun. And such being the case, it necessarily follows that the diurnal motion, too, belongs to the earth. For if the sun stood still, and the earth did not revolve upon itself but merely had the annual movement around the sun, our year would consist of no more than one day and one night; that is, six months of day and six months of night, as was remarked once previously.

See, then, how neatly the precipitous motion of each twenty-four hours is taken away from the universe, and how the fixed stars (which are so many suns) agree with our sun in enjoying perpetual rest. See also what great simplicity is to be found in this rough sketch, yielding the reasons for so many weighty phenomena in the heavenly bodies.

SAGREDO: I see this very well indeed. But just as you deduce from this simplicity a large probability of truth in this system, others may on the contrary make the opposite deduction from it. If this very ancient arrangement of the Pythagoreans is so well accommodated to the appearances, they may ask (and not unreasonably) why it has found so few followers in the course of centuries; why it has been refuted by Aristotle himself, and why even Copernicus is not having any better luck with it in these latter days.

Critics Scorned

SALVIATI: Sagredo, if you had suffered even a few times, as I have so often, from hearing the sort of follies that are designed to make the common people contumacious and unwilling to listen to this innovation (let alone assent to it), then I think your astonishment at finding so few men holding this opinion would dwindle a good deal. It seems to me that we can have little regard for imbeciles who take it as a conclusive proof in confirmation of the earth's motionlessness, holding them firmly in this belief, when they observe that they cannot dine today at Constantinople and sup in Japan, or for those who are positive that the earth is too heavy to climb up over the sun and then fall headlong back down again. There is no need to bother about such men as these, whose name is legion, or to take notice of their fooleries. Neither need we try to convert men who define by generalizing and cannot make room for distinctions, just in order to have such fellows for our company in very subtle and delicate doctrines. Besides, with all the proofs in the world what would you expect to accomplish in the minds of people who are too stupid to recognize their own limitations?

No, Sagredo, my surprise is very different from yours. You wonder that there are so few followers of the Pythagorean opinion, whereas I am astonished that there have been any up to this day who have embraced and followed it. Nor can I ever sufficiently admire the outstanding acumen of those who have taken hold of this opinion and accepted it as true; they have through sheer force of intellect done such violence to their own senses as to prefer what reason told them over that which sensible experience plainly showed them to the contrary. For the arguments against the whirling of the earth which we have already examined are very plausible, as we have seen; and the fact that the Ptolemiacs and Aristotelians and all their disciples took them to be conclusive is indeed a strong argument of their effectiveness. But the experiences which overtly contradict the annual movement are indeed so much greater in their apparent force that, I repeat, there is no limit to my astonishment when I reflect that Aristarchus and Copernicus were able to make reason so conquer sense that, in defiance of the latter, the former became mistress of their belief.

Newton Formulates Laws of Gravity and Motion

By P.M. Rattansi

In the following selection science historian P.M. Rattansi describes the discoveries that paved the way for Sir Isaac Newton's theories. German astronomer Johannes Kepler, for example, showed that the planets do not orbit in perfect circles, as had long been supposed, but rather in ellipses. Capitalizing on the work of his predecessors, Newton added the missing element: a mathematical explanation of the forces at work in the phenomenon Kepler and others observed. Rattansi explains how Newton, after contemplating the way an apple falls to the earth, worked out an explanation. Assuming that the same force that pulls the apple down also tugs on the moon, Newton hypothesized that the moon would fall toward the center of Earth at the same speed it moves at a right angle to it. These offsetting forces result in the moon remaining in a constant orbit. This insight led to what has become known as Newton's universal law of gravitation. It states that the force of gravity between two bodies is equal to the product of their masses divided by the square of the distance between them. Newton also formulated three fundamental laws of motion. Taken together, these made it possible to understand and predict the orbits of the planets with a high degree of reliability. At the time he wrote this selection, P.M. Rattansi was a professor in the Department of History and Philosophy of Science at the University of London.

P.M. Rattansi, *Isaac Newton and Gravity*. Hove, UK: Priory, 1978. Copyright © 1974 by P.M. Rattansi. Reproduced by permission.

The story of [Isaac] Newton and the apple has been told and retold. Friends heard it from his own lips in his old age. While sitting in his mother's orchard one day during the plague year, his attention was caught by an apple falling to the ground. This set him thinking, but twenty years were to pass before his ideas about gravity and motion were published in his *Principia Mathematica.*

Why the delay? A number of clever explanations have been given of why he kept his discovery from the world for so long. Recent historical work has changed our ideas about this. It now seems likely that the crucial idea of universal gravitation came to Newton only much later, not until after he had written the first section of his masterpiece.

Newton had already studied the problem of the motion of earth and planets around the sun when he was an undergraduate. Aristotle had imagined a nest of hollow spheres carrying sun, moon, and planets around the earth in their daily and yearly motions. Copernicus had not given up the spheres, but his work made others more and more reluctant to believe in them. If there were no spheres what kept the earth and the planets circling around the sun?

Kepler's Contribution

Early in the seventeenth century the German astronomer Johannes Kepler suggested that a moving force, radiating from the sun like spokes in a wheel carried the planets round as the sun itself spun round. But he altered this model after a discovery of immense importance. Kepler was inspired by Pythagoras's belief that mathematics would lay bare the hidden nature of things. He tried to show that the observed positions of the planets were on uniform circular orbits and he tried to do this more exactly than anyone else had done before him. After long years of hard work and failure, he came up with a solution. The ancient idea of uniform circular motion had to be abandoned. The planets moved in *elliptical* orbits (flattened circles) with the sun at one focus. In such an orbit the speeds of the planets and their distance from the sun would change all the time. To explain this Kepler thought of magnetic poles in the planets which were attracted or repelled by a magnetic north pole in the sun itself.

Few people took Kepler's elliptical orbits or his "celestial mechanism" seriously for a long time. A new science of motion had to be created before his ellipses, and two other discoveries, could serve as the foundation of a new world-picture. The first great step towards it was made by another follower of Copernicus, Galileo Galilei.

Galileo is often said to have disproved Aristotle by dropping balls of different weight from the Leaning Tower at Pisa. If Aristotle's theories were right, the heavier ball should have struck the ground first. Galileo showed that both balls reached the ground at the same time. The story may or may not be true. It is certainly not a good illustration of Galileo's importance to science. His great achievement was to single out the vital features of physical events in a real-life situation—like a falling stone, or a swinging pendulum—and to "think away" the rest. By doing this he clarified ideas like speed, acceleration and resistance.

Break with Ancient Thought

Aristotle had stuck close to common experience. He thought that, except for "natural" motions, all movement needed a mover. If the mover stopped pushing the movement would stop. Galileo "thought away" friction and imagined a perfectly smooth ball being placed upon an ideally smooth slope. The ball would roll faster and faster down the slope. On an upward slope, force would be needed to push it along or even to hold it still. Therefore, on an endless level surface, there is no reason why a ball once set moving should ever stop.

The principle of *inertia* to which Galileo pointed was a big break with ancient thought. Motion as such does not need a continuous force to keep it going. Only a *change* of motion needs force. Galileo used this principle to answer ancient objections to a moving earth. A stone would not cease to share the motion of the earth the moment it lost contact with it. When dropped from a high tower, it would drop to the foot of the tower because it would really have two motions: one downward, the other a circular one shared with the moving earth. The power of old ideas, even on those trying to break with them, is shown by Galileo's continuing to think of these motions as "natural." Circular motion remained perfect motion for him, only now earthy matter could share this perfection too.

It was [René] Descartes who sharply rejected such ideas. Inertial motion was uniform motion in a straight line. Circular motion was not natural, but needed a mechanical cause. For the earth and the planets, said Descartes, that cause was the whirlpool of subtle matter which carried them around the sun. That same power explained why a stone, when let go, dropped to the ground. The subtle matter in the earth's vortex moved away much faster and the stone, like a piece of wood in a watery whirlpool, was driven to the centre.

If it was the fall of an apple that set Newton thinking in 1666 of the power needed to hold the moon around the earth and the planets around the sun, he probably thought of it as due to these Cartesian vortices. When we whirl a stone in a sling, the tension in the string may seem to us a sign of the stone's tendency to fly away. It came to be called a "centrifugal," or centre-fleeing, tendency. Newton worked out a way of calculating it, and estimated what it was for the earth, moon, and planets. It then occurred to him to combine it with something that Kepler had discovered. Kepler had found that the time (T) taken by a planet to orbit the sun varied and depended on its average distance (D) from the sun in such a way that the ratio T^2/D^3 was the same for all planets. That meant, Newton discovered, that the "centrifugal" tendency of the planets must *vary inversely as the square* of their distance from the sun.

It has usually been assumed that Newton must have thought of the centrifugal tendency as balanced by a gravitational force which acts from the sun upon the earth and the planets. As that force is universal, it should be possible to test the effect of the earth's gravitational force upon the moon. The moon's inertial motion would carry it with uniform speed in a straight line. It orbits the earth because it is continuously pulled away from that path by the earth's gravitation. Newton knew that at the surface of the earth that gravitational pull causes objects, such as apples, to fall sixteen feet in the first second after starting from rest. If the pull decreases according to the inverse-square law, then at the distance of the moon its force would be 1/3600 of that on the earth. An object on the moon must then take one minute to fall the distance it fell in one second on the earth. So according to gravity, the moon should be pulled away from its straight inertial path by sixteen feet every minute. Newton tested the result by a simple calculation. He worked out the moon's acceleration, from its period of revolution and the supposed size of its orbit around the earth. He then compared the acceleration worked out in this way, with that given by the inverse-square relation. The answer agreed "pretty nearly", but not exactly, since (as we now know) he used too small a figure for the size of the earth. Recent historical work has, indeed, unearthed a moon-test by Newton, probably done in 1666, but there is no suggestion there that he had begun to think of the inward force as a gravitational attraction between sun and planets, or earth and moon.

Hooke's Stimulus

Newton put aside work on these problems, until brought back to them in 1679 by Robert Hooke, who had now become Secretary of the Royal Society. Hooke asked him to suggest problems to revive public interest in science. Hooke wrote to Newton suggesting that if a body orbited the earth so that it was *attracted* by the earth with a force varying inversely as the square of the distance, its path would be an ellipse with the earth at one focus. Hooke offered no mathematical proof. Newton later claimed that he had worked out a proof but had tossed the paper aside, "being upon other studies."

In 1684 Edmond Halley visited Newton in Cambridge. Halley had found that Hooke and Sir Christopher Wren, like himself, had guessed that Kepler's sun-planet relation meant that there is an inverse-square attraction between sun and planets. None of them could offer a mathematical proof. Hooke claimed one, but did not take up Wren's wager of a book worth forty shillings.

Halley was delighted when Newton told him not only that he had made the same discovery, but that he had proved it mathematically for elliptical orbits. He could not find the proof among his papers, but promised to reconstruct it. The result was a tract which Halley saw on a second visit some months later. Newton's mathematical genius was displayed in his solution of a problem that had baffled the others. But the crucial idea of universal gravitation, which appears in the *Principia*, was still absent from this tract. To appreciate his final achievement we must grasp the problems which faced him.

Robert Hooke saw that the principle of inertia meant that a force must be acting whenever a body moves in any way other than with uniform motion in a straight line. For example, circular motion needs a force to act continuously towards the centre of the circle. Instead of concentrating upon supposed "centrifugal" forces arising from orbital motion, Hooke pointed to the force that made a body travel in such a path. For the earth and the planets, Hooke suggested that this force was an attraction towards the centre of the sun.

Skepticism About Gravity

But a deflecting force of that sort was unacceptable to "mechanical philosophers." Galileo had rejected Kepler's ideas of an attractive force from the sun. He had even denied the old idea that the attraction of the moon causes tides on earth and he stubbornly defended his own theory that tides were caused by the earth's rotation. Descartes criticized [mathematician Gilles] Roberval for explaining the fall of heavy bodies to earth by attraction. He could not conceive of a force acting at a distance as gravitational force was supposed to do: there must be a mechanical explanation. Such ideas seemed to Galileo and Descartes to belong to the magical tendencies which had gripped the Renaissance imagination as the hold of Aristotle slackened.

Newton himself had up to now been faithful to Descartes' programme of explaining all that happened in nature by matter in motion. His acceptance of the idea of gravitational attraction plunged him into controversies to the end of his life. He could not decide whether to explain gravity as due to the direct action of God, or as due ultimately to matter in motion. What first caused him to consider such a departure from the mechanical programme?

One reason may have been the chemical and alchemical studies on which he spent so much of his time and which have continued to puzzle historians. Newton was deeply impressed by the way some chemical substances seem to have an attraction for certain others and readily combine with them. Certain other things were difficult to explain by mechanisms of the Cartesian sort. Why does matter stick together and not just fall apart? Why does water rise in thin glass tubes? Why are light rays bent away from their paths when they enter another substance, and how can flies walk on water without wetting their feet? Attractions and repulsions between particles of matter could explain all this more simply.

Was that a backward step? Not, Newton thought, if we could discover the small number of mathematical laws presumably ruling such attractions and repulsions. Matter and motion are *not* enough to explain nature. In addition, there must be certain "active principles" planted by God in nature. Without their continuing action nature would have come to a standstill long ago. Newton's scientific and religious thought came together in this view of nature which he was developing. He was deeply influenced by the thinkers known as the Cambridge Platonists. They believed that if the universe was thought of as a clockwork which God had created but then left to run itself by mechanical principles, then men would stop believing in religion. His thoughts already seem to have been moving away from purely Cartesian explanations of nature just at the time when he began once again to think about the movements of the heavens.

If the sun pulls on the planets with a gravitational force, is this a force that acts between *particular* bodies (like the specific attractions in the chemical reactions Newton had studied)? Or did *all* matter attract *all other* matter? Newton decided that the force was universal.

Distinguishing Mass

How was the pull related to the bigness or smallness of bodies? In answering that, Newton made the first clear distinction between *weight* and *mass*. The weight of a body measures the gravitational pull, which will vary with distance from the centre of the earth. The mass is the "quantity of matter" in a body. A body may weigh less at the top of a high mountain, but its mass need not have changed.

Mass and weight must be exactly proportional at a fixed distance from the centre of the earth. Kepler's law connecting the distances of the planets from the sun with the time they took to orbit it meant that if all the planets were placed at the same distance from the sun, they would move in the same orbit. Newton realized that this would happen only if the pull of the sun was exactly proportional to their very different masses. It would pull, say, four times as hard on a body four times as massive as another at the same distance.

Galileo had scandalized Aristotelians by saying that, if it were not for air resistance, all bodies falling from the same height would have the same speeds. Newton now showed that to achieve this the gravitational force would have to pull harder on a more massive body, since there was more matter to be moved. Had the strength of the pull been the same, the *lighter* body would have fallen faster. It was a surprising fact that the pull was, in fact, exactly proportional to the mass of bodies. Newton further tested the

proportionality by experimenting with pendulums, whose usefulness in dealing with problems of falling bodies had already been
seen by Galileo. Galileo had also shown that the speed of a falling
body increases regularly with time (uniform acceleration). This
could now be shown to illustrate the general principle that a constant force produces, not a constant velocity as in the old physics,
but a constant *increase* in velocity.

When calculating forces between massive bodies at great distances, like the sun and the planets, or the earth and the moon, it
was possible to treat them as "mass-points." But calculating the
forces upon, say, an apple falling to the earth, seemed to be very
complicated. Parts of the earth near at hand would pull hard, while
other parts would pull with a force that became weaker at a distance. Using his new mathematical tool of fluxions, Newton
proved that such massive and solid spheres attracted things *as if*
their mass was concentrated at their centres. In this way the problems involved were enormously simplified.

Gravitation was a "centripetal" force. A more general idea of
force was needed to create a workable science of dynamics, that
is a science of motion and moving things. That was a very tough
problem which had defeated earlier pioneers. There were so many
different ideas of forces, measured in very different ways. Newton's true genius comes across in the simplicity of his solution. All
matter possessed "inertia"—the power of resisting motion when
at rest, or changes (of direction and magnitude) when in motion.
Gravitation was an example of an "impressed" force which
changed inertial motion. Forces, said Newton, must be proportional to, and can be measured by, the changes they cause in such
inertial motion.

Three Laws of Motion

Newton set out the basic laws of his new mechanics in three laws
of motion. The first stated that every body continues in its state of
rest or of uniform motion in a straight line unless made to change
that state by forces impressed upon it. The second law related the
increase in velocity of a body to the impressed force: the acceleration gained by a body is proportional to the force and inversely
proportional to its mass. The third law stated that to every action
there is an equal and opposite reaction.

Newton now applied his new ideas to a great range of problems.

Besides free fall and impact, he analysed the much harder problems of resistance, wave motion, and the motion of fluids. He tackled the "three-body" problem, that is the problem of orbital motions where more than two bodies are involved. The motions of the earth and the planets were obviously of this sort. He was able to explain the tides, as well as the long-term motion of the axis of the earth which produced a "precession of the equinoxes" in the heavens, and predicted a bulge at the earth's equator.

Newton struck the death-blow to Cartesian vortices in the course of his investigations. He worked out that they would lose motion continuously and could not carry on for very long. Nor could any Cartesian "subtle matter" explain gravitation. The presence of an atmosphere even as thin as that of the earth in interplanetary space would make Jupiter lose one-tenth of its motion in thirty days. Instead of the full universe of Descartes, Newton's was an almost empty universe, where the emptiness throughout space and within physical objects quite overshadowed the small parcels of matter scattered through its vast reaches.

Newton Publishes

These mammoth achievements were the result of just two years' work. By April 1686 the manuscript of the *Mathematical Principles of Natural Philosophy* was before the Royal Society. Halley decided to print it at his own expense when the Society found itself unable to pay for it. Besides his financial sacrifice, Halley also had to soothe Robert Hooke, who now accused Newton of stealing the inverse-square law from him. "Philosophy," Newton sadly commented, "is such an impertinently litigious Lady, that a man had as good be engaged in lawsuits, as to have to do with her." Halley had a hard time persuading Newton not to leave out the section *System of the World* which was the crown of his work.

The great work was published in May 1687 and created an immediate sensation. Even the followers of Descartes, who rejected its central idea of an attractive gravitational force, came in time to recognize its achievement. The dream of a mathematical science of nature was at last set on firm foundations, even if at its heart lay a type of action which the founders of the mechanical philosophy had wished to banish from any rational system of science.

Einstein's Theory of Relativity Advances Astronomy

By Neil de Grasse Tyson

Albert Einstein's theory of relativity has proven of great value to astronomers in understanding time, space, light, and gravity. In 1905
Einstein proposed that gravity, like acceleration, can warp time and
space. Astronomer Neil de Grasse Tyson recounts in the following
selection how Einstein's theory was first proven correct and explains
its current usefulness to astronomers. In 1919 British astrophysicist
Arthur Eddington set out to test whether the sun's gravity would
bend starlight, as Einstein's theory predicts. Taking advantage of a
solar eclipse, he and his team recorded the apparent position of stars
near the sun. Six months later, when Earth had traveled to the opposite side of the sun, they recorded them again. Eddington found that
the sun's gravity had shifted the apparent position of the stars, just
as Einstein's theory had predicted. Since then, Tyson explains, astronomers have applied knowledge of gravity's light-bending effect
to make sense of their astronomical observations. Neil de Grasse
Tyson is the Frederick P. Rose Director of New York City's Hayden
Planetarium. He also teaches astrophysics at Princeton University.

One of the most mind-bending discoveries of twentieth-
century astrophysics—predicted in 1911 by Albert Einstein and incorporated into his general theory of relativity in 1915—is that matter curves the fabric of space. More

Neil de Grasse Tyson, "Darkness Visible," *Natural History,* vol. 106, February 1, 1997, p. 76.

recently, the phenomenon has proved to have applications that are indispensable for probing dark matter in the universe.

But why should anybody believe that matter curves space? Support for the idea does not come from wishful thinking but from experiment. In what is now considered a classic test of general relativity, one can take advantage of a total solar eclipse to prove that starlight is bent in the curved space of the Sun's gravity. This measurement can be made only during an eclipse because, of course, stars are not otherwise visible from Earth during the daytime.

Contrary to popular belief, total solar eclipses are not rare. On average, one takes place somewhere on Earth's surface every one and a half years, although a particular location can go several hundred years without one. The perceived paucity derives from the narrowness of the eclipse path across Earth's surface. Opportunities to test Einstein's idea came immediately with the solar eclipses of 1916 and 1918, but both of these were unfavorable for several reasons, including a dearth of bright stars near the Sun's edge during totality. In addition, the First World War was raging, hindering the dissemination and digestion of general relativity, and making a safe and reliable eclipse expedition unfeasible.

By 1919, however, all was quiet, and the English astrophysicist Arthur Stanley Eddington mounted well-publicized dual expeditions (as a hedge against bad weather) to view the May 29 total solar eclipse from South America and from Africa. Eddington's goal was simple: to map the precise positions of stars in the vicinity of the Sun. The complete experiment, however, required that he map the same star field six months later, when Earth was on the other side of its orbit and the Sun's gravity safely out of the way. Only when the two images were compared could he reliably deduce whether the stellar positions had indeed changed.

Eddington Proves Einstein Right

To the world's astonishment, Eddington found a shift in stellar positions that agreed with Einstein's prediction of 1.75 arc seconds to within experimental accuracy. (The shift is an angle smaller than the thickness of a dime when viewed over the full length of a football field.) Einstein became an immediate celebrity. Actually, with hindsight we could have predicted that a light path would bend in response to gravity, using simple Newtonian laws and the equivalence of mass and energy. Isaac Newton himself had suspected such a thing. At the end of the second edition (1717) of his seminal treatise on optics, Newton presents the reader with some unsolved mysteries:

> I shall conclude with proposing only some Queries, [for] a further search to be made by others.
>
> Query 1. Do not bodies act upon light at a distance, and by their action bend its rays; and is not this action strongest at the least distance?

But Newton's equations alone will give you the wrong answer— only half of what was predicted by Einstein's equations. Why? In relativity theory, where space and time are conjoined, time itself can also be thought of as bending in the presence of gravity—a concept for which there is no analogue in pre–twentieth-century physics. Light therefore takes slightly longer to pass the Sun than it otherwise would, which serves to increase the angle of deflection. We should thus speak not of the curvature of space but of the curvature of space-time.

The geometric ingredients are simple: all you need is an observer, a distant source of light, and a massive object that falls somewhere along, or close to, the observer's line of sight. In 1936, Einstein imagined a case in which two stars were perfectly aligned, with the background star serving as the light source and the foreground star serving as the source of curvature. In this layout, light paths can bend not only around to one side but also around to the other side. They can also bend above and below. Indeed, since space-time is curved everywhere in the foreground star's vicinity, perfect alignment will force the light to fan out into a complete "Einstein ring."

Perfect alignments on the sky are rare, but the varieties of light sources and gravity sources that produce visible bending are practically limitless. Because the action of curved space-time upon light greatly resembles the action of ordinary optical lenses, the phenomenon is known as gravitational lensing. A colleague of mine has even forged a career in mathematics by exploring the abstract theory of gravitational lenses—an elegant field in which one seeks to describe all possible lenses, alone and in combination, regardless of whether an example in the real universe is known or will ever be found. But there's no need to be jealous. Even if we tally only the more common lens configurations, we retain a veritable fun house of images.

The action of a gravitational lens is not restricted to visible light. If light takes the shortest distance through space-time between two points (which it does), and if space-time itself is curved (which it is), then the path of light will curve along with it, regardless of whether the light is composed of gamma rays, X-rays, ultraviolet, infrared, microwaves, or radio waves. And, most importantly, the cosmic stuff that causes the bending can be made of absolutely anything, as long as it contains enough mass to curve space-time measurably in its vicinity.

Apart from the Sun, the first bona fide gravitational lens was discovered in 1979, when a suspicious-looking double quasar was found. Its two images were much, much closer together over the 41,253 square degrees of the sky than we would expect as a result of mere chance. Quasars are extremely distant: ever since the first quasar was discovered in the early 1960s, the farthest known objects in the universe have always been quasars (although today, the farthest known galaxies are more distant than the nearest quasars). They are also extremely luminous: many are hundreds of times the luminosity of all hundred billion stars of our Milky Way galaxy. And finally, they are extremely small, generating nearly all their energy in a region no bigger than our solar system. But apart from these properties, quasars have little in common.

An analysis of their light, spread into its component colors, shows quasars to be as individual as fingerprints, yet the two images of the suspicious-looking double quasar were identical. The choice was clear. Either they were two separate, identical quasars that happened to be insanely close to each other on the sky and in space or they were a single quasar whose light was split into two images by the gravity of a faint object in the foreground. The lens explanation won handily. And several dozen other gravitationally lensed quasars have been discovered since.

The theory of gravitational lenses tells us that an odd number of images is always formed—the single image that passes straight through the center of the lens, plus split images falling elsewhere. The only exception is the case of perfect alignment, which produces the Einstein ring. Usually, however, the lens contains a blob of gaseous matter that absorbs light from the central image, rendering it barely visible. The fun part comes when, depending on the shape of the lens, the background source of light increases in size, magnifies in brightness, inverts, and gets all bent out of shape.

Quasars are, of course, excellent image candidates because their extreme distances grant them the best chance of having something fall between us and them—commonly a foreground galaxy whose curved space-time can be quite complicated: often there is a black hole in its nucleus, a dense central bulge of a few billion stars, a disk, and a halo of dark matter that extends well beyond the visible region of the galaxy. In this layout, image pairs can show up almost anywhere.

Huchra's Lens

A famous example is Huchra's lens, named for John Huchra, the Harvard astrophysicist who discovered the first quartet of identical quasar images that were formed in the gravity field of a foreground galaxy. (A fifth, central quasar image, as expected, is hidden behind the lensing galaxy.) More generally known as Einstein's crosses, the few known examples of quasar quartets have all been imaged by the orbiting Hubble Space Telescope. These lens antics are not just intellectual curiosities—their exact locations allow us to probe the distribution of matter in the intervening galaxy. The split among the images is so tiny, however, that an ordinary groundbased telescope looking through Earth's atmosphere will see one fused blob.

As carpenters know well, what you really want in life is a useful tool. A clever use of geometry allows us to use lensed quasars as a tool to derive the famous Hubble constant, and thus the expansion rate of the universe. If the quasar happens to flare abruptly (as many quasars do on occasion), and if the pathways for each image differ slightly from one another in length, then the brightening will reach Earth at different times. Knowing the speed of light and the time delay among the images, we can use the geometry of the lens (inferred from a model of the mass that does the bending) to derive the Hubble constant. Recent estimates made with this approach fall right in the range determined using other methods.

Another application uses a low-mass star or a Jupiter-size planet as a lens to a background star. At great distances, both low-mass stars and Jupiter-size planets are much too faint to be detected. How can we know if these galactic vagabonds are out there? And if they are, do they dominate the mass of the galaxy? Such a feeble lens would produce so narrow a split that multiple images would be imperceptible, a phenomenon known as "microlensing." Because we can't resolve them, these multiple images appear as a single blob. Sounds hopeless. But in 1986, Bohdan Paczynski, a Princeton astrophysicist, noticed that the random motions of objects in our galaxy allows you to watch the brightness of a background star increase dramatically and then, as a foreground lens passes by, decrease back to its original level over a week or two.

Sounds simple. But to catch such an event in our galaxy, you would have to watch a given star for a quarter million years. Nobody really wants to wait that long, so we do the next best thing. We simultaneously monitor the brightnesses of a million stars every night of the year. Every twelve months we can capture anywhere from two to six events, provided we are not confused by the tens of thousands of "variable" stars, stars that naturally and continuously oscillate in brightness. Following up on Paczynski's prediction, several groups—including one led by Paczynski himself—launched major surveys of our galaxy to detect these microlensing events. Paczynski's group looked through the galactic plane toward the bulge and found lenses in just the numbers expected. They also found cases in which the lens passed across an unresolved double star, which gave two brightenings in rapid succession with no activity either before or afterward.

Identifying Dark Matter

The total number of recorded microlensing events is still meager. As the numbers improve, however, we may be able to classify these invisible lenses and thus ask questions about dark matter in our galaxy: How many isolated Jupiter-size planets are out there traveling among the stars? How many stars are too faint to see? How many dead stars are there? How about black holes? Might there be more mass locked up in dark objects than in luminous objects? Answers are forthcoming.

In another microlensing survey of our galaxy, a team based at Berkeley [California] and Canberra, Australia, has probed the spherical halo of our galaxy (known by other methods to contain large quantities of dark matter) by looking for lensing events against the individual stars of the Large Magellanic Cloud, a "background" galaxy that is close enough to the Milky Way for many of its individual stars to be seen. Preliminary, though controversial, results suggest that much of the dark matter in our halo (and by extrapolation, other galaxy halos) may be composed of white dwarfs—extremely dim stellar corpses.

All stars, and nearly all quasars, are so small on the sky that when photographed from Earth, lensed or otherwise, they appear as simple points of light, which lend themselves well to microlensing and simple split quasar images. But for sheer visual splendor, the combination of a background galaxy lensed by a foreground cluster of galaxies wins all contests. Each one of the hundreds of galaxies in a massive cluster can act as a lens, as can the collective gravity of the entire cluster—dark matter and all. In the end, what you get are many pathways, some of them rather exotic. Typically, the background galaxy's images are curved into arcs that are each concentric with the cluster's center of mass. Sometimes these titanic cosmic mirages prove to be galaxies that would have gone undetected without the magnifying effect of the curved space-time in the foreground cluster. Indeed, we have discovered a peephole to the distant universe.

The brightnesses, shapes, and locations of the multiple images of the background galaxy can actually be used to solve the lens problem in reverse: What must be the structure of matter in the lensing cluster to have created the observed assortment of lensed images? Sounds complicated. It is. And it especially excites my aforementioned colleague, the lens mathematician. As already suspected, more than 90 percent of the mass in galaxy clusters is invisible and remains unknown. The triumph of our understanding of curved space has brought us face to face with a gaping area of cosmic ignorance. Looks like we need more tools.

Hubble Discovers the Expanding Universe

By Gale E. Christianson

Edwin Hubble (1889–1953) made a discovery that changed humanity's view of the universe. In the following selection Gale E. Christianson explains that in 1929 Hubble showed that galaxies are rushing away from Earth. Working with images captured by his assistant Milton Humason through the largest telescope then available, the Mount Wilson Observatory's 100-inch reflector, Hubble measured the redshift, or stretching, of light from galaxies. From these photographs Hubble concluded that the farther away a galaxy, the faster it is receding. This theory has become known as Hubble's law. With painstaking care Hubble and Humason spent countless hours making observations and checking calculations. Through their work Hubble and Humason provided the first clear evidence that the universe is expanding. Author Gale E. Christianson has written several biographies of major scientific figures as well as a book on global warming.

[E]dwin] Hubble's research on Cepheids [stars with a standard brightness] had provided him with a wealth of distance measurements, but he was far from content. He thought, spoke, dreamed, and wrote of his work to date as mere "reconaissance." The 100-odd Messier objects [fuzzy objects] were as familiar to him as the alphabet, and he knew his own nebula (the Milky Way), with its complicated structure of bright and dark nebulosities, star clusters, and planetaries, as thoroughly as any pilot feeling his way through a treacherous system of channels, rip currents, and shoals. But the Indies, supposedly somewhere far off in the distance, were yet to be sighted. In charting deep space, the period-luminosity relation for Cepheids and other stellar bodies would have to be exploited, as never before. Thus he continued his work on distances while [astronomer Milton] Humason readied the telescope for the task of photographing the spectra of the nebulae. If, as Hubble suspected, a nebula's velocity of recession was truly an index of its distance, then the distances of nebulae far across the universe could be inferred by simply measuring their redshifts [caused by the stretching of light from receding galaxies, which makes the light appear redder than normal]. Within weeks the unlikely pair hoisted sail and set a course for far-flung waters, Humason in the crow's nest and Hubble at the helm.

Poor but memorable was how Humason characterized his first plate. He intentionally picked a nebula whose redshift [astronomer Vesto] Slipher had not been able to obtain because of its distance from Earth. Photographing through a yellow prism which blocked the ultraviolet light, the assistant astronomer passed two frigid nights awaiting the result. He then developed the plate, and, with the aid of a magnifier, quickly located the so-called H and K lines produced by calcium atoms in the nebula. Though the spectrum was faint, the telltale vertical marks were shifted to the right or red end, as had been expected.

Humason immediately telephoned an elated Hubble, who was waiting for him at Santa Barbara Street when he finished his run. Hubble confirmed the redshift, calculating it at some 3,000 kilometers per second, about 1,800 kilometers per second greater than Slipher's highest value. When he was queried about his feelings at the moment of triumph years afterward, an educationally impoverished Humason claimed [poet Lord] Byron as his model, though he could offer no reason as to why.

Speeding Up the Process

Yet Humason's "adventures among the clusters," as Hubble characterized them, were nearly thwarted by his own mutiny. When pressed to do more, he balked. The "tremendously long exposures" hardly seemed worth the pain and suffering exacted by the mountain.

At this point, Humason received a call from [astronomer George] Hale, requesting him to stop by the solar observatory he had constructed in Pasadena [California] after giving up the directorship of Mount Wilson. Having been briefed by Hubble—or perhaps by [astronomer Walter] Adams on Hubble's behalf—Hale asked his visitor to continue with the project, promising him all the technical support he required, including a faster spectrograph and an improved camera. Humason, who was already beholden to the great man for promoting him to assistant astronomer despite deep misgivings, was both flattered and touched. After discussing a few more technical details, he agreed to push ahead.

The promised spectrograph and camera were designed by J.A. Anderson of the physical laboratory and constructed in the observatory shop. Humason described the new equipment as "very fast—at least we thought it was fast compared to what I had worked with before." An exposure that had taken two or three nights to obtain was now ready in a few hours.

Assuming the role played by a lonely Slipher over the years, Humason began by gathering spectra on many of the same forty-five nebulae photographed by his predecessor atop a pine-capped peak near Flagstaff [Arizona]—great spirals like M31, M33, M51, M101—the heart of [French astronomer Charles] Messier's catalogue [of nebulae], compiled a century and a half earlier. Not surprisingly, the redshifts were confirmed: in all directions, the nebulae appeared to be moving away from Earth, or Earth from them. On the basis of his calibration work with Cepheids, Hubble quickly established the first linear relation between the degree of spectral displacement and the estimated distance to the observed object— the greater the redshift, the more remote the source of light.

Caution prevailed. He wrote [astronomer Harlow] Shapley in May 1929 that his recently published paper, "A Relation Between Distance and Radial Velocity Among Extra-Galactic Nebulae," had been held "for over a year." He had wanted to wait even longer pending the accumulation of further data on fainter nebulae, "but we knew from past experience that others would rush into print the moment the new large velocities were known." His distasteful confrontation with [astronomer Knut] Lundmark over the classification of nebulae was still weighing heavily on his mind.

The Swedish astronomer had recently written Adams seeking permission to return to Mount Wilson for the purpose of "determining radial velocities and the spectrographic rotation of spiral nebulae." If possible, he would like to draw on the technical expertise of Milton Humason. "As you might have heard, Dr. Hub-

ble and I have agreed this summer [in Leiden] to get into better
mutual understanding." Wasting no time, Adams cabled Lundmark, who had recently become director of the Observatory of
Lund: "Cannot give you observing time with telescopes. Nebular
program already arranged." In a follow-up letter Adams left no
doubt that the Hubble-Lundmark Fault was still active, if only
from the American side: "Dr. Hubble and some of the rest of us
have been laying out for some time past a program of work on the
radial velocities of spirals. . . . For this reason I do not feel that it
would be desirable for you to undertake similar work at this observatory." Should Lundmark still wish to make the journey he
was welcome to use the photographic archives.

A Groundbreaking Publication

Though only six pages in length, Hubble's first paper on the
velocity-distance relation represented a giant step in modern cosmology. Writing of Hubble for the *Dictionary of Scientific Biog-
raphy*, the noted cosmologist G.J. Whitrow asserted that he had
wrought as great a change in humankind's conception of the universe as the Copernican revolution four hundred years before. In
place of a static picture of the cosmos, it seemed to many that the
universe must be regarded as expanding, the rate of the mutual recession of its parts increasing with their relative distance. Yet
nowhere in his modestly drafted paper did the astronomer mention the expansion of the universe or, for that matter, the universe
itself. "New data to be expected in the near future," Hubble wrote
in conclusion, "may modify the significance of the present investigation or, if confirmatory, will lead to a solution having many
times the weight."

Truth to tell, Hubble was already in possession of large amounts
of new data and pushing hard for more. Perched, like a monkey,
on the small Cassegrain [viewing] platform five stories above the
observatory floor, his face grotesquely illuminated by red dark-
vision lamps, a freezing Humason prodded and coaxed the reluctant beast through moonless nights, punctuated by staccato winds
and the incessant ticking of the weight-driven clock. Swallowed
up by the dark, the astronomer's presence was periodically signaled by chimes, indicating that he had pressed the button changing the drive rate to keep his guide star in place. If the mechanism
balked, which happened all too frequently, he held the image in

place by forcing his shoulder against the great cannon, and occasionally climbed onto its iron frame, bending his body at painfully awkward angles for the sake of the embryonic plate steeping in light from a nebula time out of mind. "You had to stretch out into nothing," he recalled, but sometimes even this was not enough. A whitish foam accumulated on top of the mercury in the float tanks, multiplying and roiling like a witch's brew until it set up an uncontrollable vibration. When this happened, the telescope had to be shut down while the troubled "waters" were stilled by skimming the fulminate from the vats.

One week after Hubble informed Shapley of the need for more data, an exuberant Adams wrote Carnegie Institution president John C. Merriam, "We are getting some amazing results on the spectra of very distant spiral nebulae." Photographing in the great cluster centered in the constellation Virgo, Humason and Francis Pease, who is perhaps best remembered for his Mount Wilson collaborations with Albert A. Michelson, the Nobel laureate physicist, had obtained displacements corresponding to velocities ranging from 3,500 to 8,000 kilometers per second. Having more than quadrupled Slipher's largest displacement, Humason was adding new velocities at the rate of ten a month, forging beyond Virgo to Pegasus, Pisces, Cancer, Perseus, Coma, and Leo, whose speed of recession was calculated at a staggering 19,700 kilometers per second.

Celebrating Recession

"Nick" Mayall, a graduate student at Berkeley who was working on the mountain, happened to be with Humason when he phoned Hubble from the dome informing him of the nearly 20,000 kilometers per second redshift, which Hubble himself had predicted. "I was so close to the phone I heard him tell Milt, 'Now you are beginning to use the 100-inch the way it should be used.'" Humason uncorked a bottle of his Panther Juice and they toasted the previous night's success. After gaining eminence as an astronomer in his own right, Mayall looked back on this period as "the most stimulating of my life."

Leaving as little as possible to chance, Hubble kept strict tabs on Humason's schedule.

When I got back from the mountain, he would come striding down

the hall to ask what luck I had had. . . . He was very fast in mathematical calculations. When I brought him new exposures of spectra he would pick up a pencil and a pad, and jot down figures as fast
as the pencil could move, and have the distances in a few minutes.

If Humason was successful, which was most of the time, he knew
what was coming next. The procedure must be repeated in order to
obtain an independent check on the first result. In using the term
"weight" in the conclusion of his initial paper on redshifts, Hubble
meant exactly what he said, the mind of the skeptical scientist echo-
ing the teachings of the ancient prophet Isaiah, which he read in
front of a crackling fire: "Precept upon precept, precept upon precept; line upon line, line upon line; here a little, and there a little."

Comparing Galaxial Brightness

The Cepheids, novae, and blue stars that had first guided Hubble
deep into the universe were of little use in determining distances
now. In the realms he plied with Humason, whole nebulae (galaxies) contained in giant clusters such as Ursa Major and Boötes
were invisible to the eye, even through the 100-inch. Believing it
reasonable to assume that clusters of nebulae are similar to one
another, as individual nebulae are, he compared the brightest stars
in the largest nebulae of the Virgo cluster with the brightest stars
in the Milky Way. The idea of comparing the apparent brightness
of two objects thought to have the same true brightness was then
applied to the nebulae themselves. The most prominent members
of the Virgo cluster turned out to have about the same true brightness as the Andromeda nebula. Once this was established, Hubble could compare a prominent member of any distant cluster with
a prominent member of the Virgo cluster, just as he had compared
Cepheids in the past. If the apparent brightness of the distant nebulae was one hundred times fainter than its Virgo counterpart,
Hubble calculated it to be ten times farther away, having already
demonstrated that apparent brightness decreases with the square
of the distance. Soon, as Mayall had witnessed, Hubble was able
to estimate the redshift Humason was after before the spectrum of
the cluster nebulae could be captured on a plate.

  "The Velocity-Distance Relation Among Extra-Galactic Nebulae," co-authored with Milton L. Humason, appeared in the year's
first issue of *The Astrophysical Journal*, in 1931. What Hubble had

termed the "rather sketchy" data contained in his 1929 paper was fortified by Humason's redshift measurements of another fifty nebulae, thirty-one of which occurred in clusters. The startling figures on the constellation Leo commanded the most attention: one of its clusters was receding from Earth at nearly 20,000 kilometers per second, placing the faint object 105 million light-years away, each light-year representing some 6 trillion miles. Amazing though it seemed, Hubble had every reason to believe that this sudden expansion of the cosmos was more the beginning of the journey than the end. He predicted that redshifts corresponding to distances at least three times as great were within the realm of possibility, and included a table with which to support his hypothesis. This numbing realization added to the already considerable fear that Mount Wilson would run out of telescope before God ran out of universe. In July 1927, an article in the *Los Angeles Illustrated Daily News* had made known Hale's recurring dream of constructing a 300-inch behemoth, "the housing for which would stand taller than the Statue of Liberty." No less an enthusiast, Hubble was also quoted: "The $12,000,000 instrument has been declared possible and the subscription of funds for its erection is anticipated."

Only tentatively developed in the late twenties, the now famous principle destined to bear Hubble's name appeared full-blown in his most recent article. According to *Hubble's law* or the *law of redshifts*, the distances and recessional speeds of nebulae are in direct proportion to each other. Double the distance to a nebula and the speed doubles; triple the distance and the speed triples. A nebula at a distance of 100 million light-years is moving away from Earth twice as fast as a nebula that is only 50 million light-years away. Thus, as Humason was destined to do many hundreds of times before his retirement, an astronomer takes a spectrum of a galaxy whose individual stars and globular clusters are too faint to be observed, determines its speed by means of the redshift, then plots its distance on a graph encompassing Hubble's elegant construct—the speed in thousands of kilometers per second on the vertical axis, the distance in millions of light-years on the horizontal axis.

The Basis of Hubble's Law

Yet to perform what would soon become a routine function, Hubble had to know the velocity or what he cautiously termed the "apparent velocity" of the expansion rate. His calibration of the distance to Cepheid variables and certain of the brightest gaseous nebulae led to the formulation of the K term. In 1929, he had calculated that for every million parsecs (1 parsec = 3.258 light-years), a nebula is receding at an additional 500 kilometers per second. By 1931, with the accumulation of more data, he revised the K term upward, calculating it at 558 kilometers per million parsecs. In time, astronomers would replace the K with the letter H, naming the value for the expansion rate *Hubble's constant.* Hence the formula used to calculate the velocity-distance relation, *Hubble's law*, is simply expressed: $V = Hd$.

As he had done in each of his previous papers, Hubble handled the theoretical implications of his findings with what one chronicler has called "long tongs." "The present contribution concerns a correlation of empirical data of observation," the reserved astronomer wrote at the end of some forty revolutionary pages. "The writers are constrained to describe the 'apparent velocity-displacements' without venturing on the interpretation and its cosmologic significance." He had taught the man assisting him equally well. Long after Hubble was gone and their joint labors had shaken modern cosmology to its foundations, Humason told an interviewer:

> I have always been rather happy that my end of—my part in the work—was, you might say, fundamental, it can never be changed—no matter what the decision is as to what it means. Those lines are always where I measured them and the velocities, if you want to call them that or red shifts or whatever they are going to be called eventually, will always remain the same.

Radio Astronomy Unveils New Wonders

By Michael Rowan-Robinson

Astronomy normally brings to mind a person peering through a telescope at some distant planet or star. Radio might make one think of broadcast towers. Since the 1930s, however, astronomers have learned a lot by capturing radio waves emitted by various celestial objects. In the following selection astronomer Michael Rowan-Robinson describes the development of radio astronomy. The story begins, he writes, with a Bell Labs researcher, who, in the early 1930s, was trying to locate the sources of static that interfered with radio broadcasts. Karl Jansky, the researcher, found that one of those sources came from far beyond Earth's atmosphere, in the vicinity of the constellation Sagittarius. Jansky was the first to show that celestial objects emit radio waves that can be detected on Earth. Within a few decades radio astronomy had become an important new branch of astronomy. Rowan-Robinson explains that by pointing large radio-wave receiving dishes at the sky, astronomers have discerned new details about the sun and Jupiter. They have also been able to detect many celestial objects that are not visible in ordinary light. Among these are interstellar gas clouds, certain features of galaxies, and rapidly spinning, shrunken stars called pulsars. Michael Rowan-Robinson is head of astrophysics at Imperial College in London. His research interests include the study of distant galaxies.

Instead of the eye or the photographic plate, we detect radio waves with specially designed radio receivers. Radio waves were first discovered by Heinrich Hertz in 1888 and first used practi-

Michael Rowan-Robinson, *Cosmic Landscape: Voyages Back Along the Photon's Track.* London: Oxford University Press, 1979. Copyright © 1979 by Michael Rowan-Robinson. Reproduced by permission of the author.

cally by Guglielmo Marconi in England in 1895. Their break-throughs followed the sensational recognition by James Clerk Maxwell, in 1864, that light consisted of nothing more than an oscillating electro-magnetic field. At any point in space an electric charge will be pushed in the direction of the local electric field, if there is one. A magnet will line itself up in the direction of the prevailing magnetic force field. The earth itself acts like a huge bar magnet and at any point on or near earth's surface, a compass needle, which is nothing more than a small magnet, will align itself in the direction of the earth's magnetic field. Now when a light wave passes all that we notice is a trembling of the electric and magnetic fields and then stillness again. That trembling is the light wave.

The wave pattern travels along at the speed of light and like the waves on the sea it carries energy with it. This energy can illuminate our visible landscape if the waves have visible wavelengths, warm us if they have infrared wavelengths, or make a radio receiver work if they have radio wavelengths. When we try to make oscillating magnetic and electric fields in the laboratory, it is natural that our apparatus will tend to have a scale of centimetres or metres, which is why the first efforts to synthesize light led to the discovery of a new kind of light, radio.

How does a radio work? When radio waves fall on a long piece of wire connected to the ground, a tiny electric current flows back and forth along the wire in time with the wave, driven by the varying electric field of the wave. The wire will respond much more strongly to radio waves of wavelength about the length of the wire. To tap this electrical energy we need to connect to the wire an electrical circuit with a 'rectifier' in it. A rectifier is a device which only allows the current to flow in one direction along the circuit. We then use this current to power a pair of headphones. The simple 'crystal' set, which gives a very adequate performance in the medium waveband, is not much more complicated than this. A long piece of wire, the 'aerial', is connected to the ground to respond to the radio waves. The current flowing to and fro along the aerial is drawn off into a simple electrical circuit which can be 'tuned' to respond to waves of only one particular wavelength. The rectification is provided by a silicon or germanium crystal diode. Your pocket transistor set elaborates this circuit by replacing the crystal with a transistor which also amplifies the signal crossing it and by providing subsequent stages of amplification to feed a loudspeaker. The beauty of the crystal set is that it powers the headphones directly with the energy of the incoming radio wave, whereas transistors or other types of diode require a source of electricity.

The receivers of radio astronomy are related to the pocket transistor much as a Formula One racing car is related to the family car: they are real specialist jobs, very sensitive, and highly tuned. They have to be to filter out the local human din and pick up the whispers from the cosmic landscape.

The Origin of Radio Astronomy

The birth of radio astronomy was also the birth of the new astronomy, the astronomy of the invisible wavelengths. . . . It is true that ground-based astronomy had long pushed out in wavelength a little way beyond the visible band, into the near infrared and near ultraviolet, but the explosion of knowledge in these bands was to follow that in the radio band. The new astronomy began quietly with a routine investigation of radio 'static' interference at a wavelength of 15 metres by Karl Jansky, at the Bell Telephone Laboratories during 1930–3. He built a rotatable aerial array 30 metres long and 4 metres high, mounted on four wheels taken from a Model T Ford, nicknamed the 'merry-go-round'. He found that part of the static was due to thunderstorms; intermittent crashes due to near-by ones and a steadier weaker noise due to many distant storms. But there remained a steady weak hiss from a direction which moved around the sky a little each day. This direction turned out to be the constellation of Sagittarius. Jansky had measured radio emission from the Milky Way.

Few astronomers took any notice of his results and Jansky himself had to return to his normal telecommunication assignments. The development of radio astronomy would have come to a halt but for a remarkable character, the American amateur astronomer Grote Reber. Fascinated by Jansky's discovery he decided to build a radio telescope in his back garden. To collect the radio waves more efficiently he constructed a large metallic dish in the shape of a parabola. This reflected the radio waves and focused them onto his aerial. With this magnificent 31-foot parabolic reflector, which must have amazed his neighbours, he used to observe the radio sky between midnight and 6 A.M., before driving to work for a radio company during the day. Between 1938 and 1944 he produced an admirable radio map of the Milky Way at a wavelength of 1.87 metres. American astronomy, dominated by the optical astronomy made possible by the large optical telescopes, failed at first to build on the work of these two pioneers.

Britain's Contributions

Meanwhile in wartime Britain, the third and most influential of the three pioneers of radio astronomy, J.H. Hey, had started the investigation which led to several astronomical discoveries and to the formation of radio astronomy as a branch of science. Neither an astronomer nor a radiophysicist, he had taken a job in 1942 as leader of a small group with the task of investigating enemy radar jamming for the army. The enemy turned out to be the sun. (At Bell Telephone Laboratories, G.C. Southworth independently detected radio emission from the sun during 1942.) After the end of the war, Hey's group discovered radio emission from meteor trails and, far more significantly, the first discrete radio source. This was Cygnus A, in the constellation of Cygnus, the Swan.

When the first sources of radio emission distinct from the general radiation of the Milky Way were discovered, their positions were so poorly known that they were just given the name of the constellation they happened to be in the direction of, with a letter. Other prominent radio sources found soon after Cygnus A were called Cassiopeia A, Taurus A (this turned out to be the Crab nebula), and Virgo A (which was in fact M87).

The work of Hey's group led directly to the formation of the two major British radio observatories. Bernard Lovell and Martin Ryle had both worked on airborne radar during the war. Afterwards they returned to their universities, Manchester and Cambridge, and started work on radio astronomy with the receivers and aerials no longer needed for radar. Lovell worked on the radar echoes from meteor trails, with some initial assistance from Hey, and later conceived the idea of the 250-foot Jodrell Bank telescope in order to detect radio echoes from cosmic rays. Ryle decided to observe the sun in more detail and started to develop 'interferometer' radio telescopes, where the output from two separate small dishes is combined to give the same ability to see fine detail, or 'resolving' power, as a large dish.

In Australia the development of radio astronomy also began during these post-war years, under J.L. Pawsey, and their first observations were also of the sun, using a different type of interfer-

ometer to Ryle. Radio waves reflected from the sea were combined with those observed directly by a cliff-top radio telescope.
Thus began the rivalry between Cambridge and Australian radio
astronomers which was to continue for many years.

Great Advances

The thirty years since the work of Hey's group have seen phenom-
enal progress, nowhere more strikingly than in the telescopes
themselves. They are among the most beautiful engineering cre-
ations of these decades. Jodrell Bank and Cambridge, England;
Parkes, Australia; Owens Valley and Hat Creek, California; Green
Bank, West Virginia; Westerbork, Netherlands; Medocina, Italy;
Nancay, France; Effelberg, Germany; Arecibo, Puerto Rico; Ooty,
India; Soccoro, New Mexico. Any of these places, and many others, are worth a pilgrimage to see the delicate traceries of metal
gazing at the sky. They have to be so large because what counts

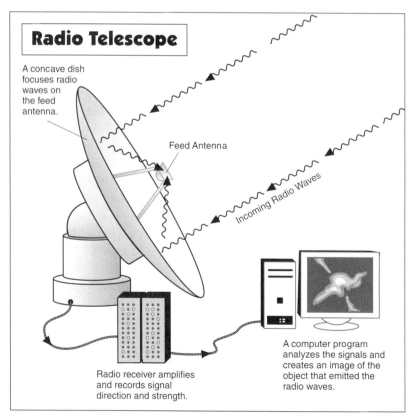

Radio Telescope

A concave dish focuses radio waves on the feed antenna.

Feed Antenna

Incoming Radio Waves

Radio receiver amplifies and records signal direction and strength.

A computer program analyzes the signals and creates an image of the object that emitted the radio waves.

when you are trying to resolve small detail in the landscape is how big your light collector is in terms of the wavelength you are using. The pupil of the eye is about six thousand visual wavelengths across and can distinguish two points separated by about one minute of arc, which is the angle made by a sixpence or dime about half a mile away. At 178 MHz, or 1.7 metres wavelength in the radio band, a similar resolving power would need a telescope 10 kilometres across. A single dish this size would be too unwieldy and colossally expensive so radio astronomers use several much smaller dishes strung out in a long line. When the signals from these are combined and the earth's rotation is used, a very large telescope can be laboriously simulated. The largest fully steerable dish is at Effelberg and is 100 metres in diameter. At Arecibo a natural valley has been used to make a fixed dish 305 metres across which can observe objects as they cross above it.

Limits of Radio Astronomy

Not all radio frequencies from the cosmic landscape reach the earth. The lowest frequencies (longest wavelengths) do not penetrate to the earth, for above the earth's atmosphere is a layer of electrically charged atomic particles (ions and electrons) called the 'ionosphere' which reflects low-frequency waves away. It also incidentally assists long-wave communication by reflecting mankind's transmissions back to earth. This is how radio waves, which travel in straight lines, can be sent round the curved earth.

The ionosphere introduces us to an important state of matter not familiar to us in the terrestrial environment, ionized gas. This is gas where the atoms have lost one or more of their electrons. This can happen through collisions between atoms when the gas is very hot or because the atoms of the gas have been bombarded by very energetic photons. The loss of the electrons, which carry a negative charge, leaves the residual atoms with a positive charge. They are called 'ions' in this state. The positive ions and the negative electrons rush around freely in the gas and respond to any electric field that is present. An ionized gas does not allow low-frequency radiation through at all because the ponderously varying electric field of the electromagnetic wave is immediately neutralized by the rapid motions of the ions and electrons. The free electrons in the gas also act as scatterers of radiation, but how effective this scattering is depends on how much gas there is.

As well as absorbing and scattering light, an ionized gas is also a source of radiation through a mechanism called 'free–free' radiation. A free electron rushing around through the gas sometimes passes rather close to an ion. When it does so it comes under the influence of the electrical attraction of the ion (unlike charges attract each other). During this period of close encounter with the ion, the electron loses some energy in the form of photons. This radiation emerges from the gas with a very wide range of frequencies. It extends all the way from the radio band to the visible band, and to higher frequencies still if the gas is very hot. This is the signature of an ionized gas cloud if it is not so dense and opaque that we cannot see right through it. If the cloud does become very dense and opaque, the distribution of radiated energy with wavelength (the spectrum of the cloud) goes over to that of a hot body at the temperature of the gas (a thermal spectrum).

One of the few examples of ionized gas on earth is the track of a lightning flash. The discharge of electricity from a thundercloud to the ground ionizes the air all along its path for a brief instant. The sun is an ionized gas cloud and in fact much of the matter in the universe is probably in an ionized state. The earth's ionosphere, mentioned above as a barrier to radio astronomy, is formed by ultraviolet and X-ray photons from the sun bombarding the top of the earth's atmosphere, stripping electrons from atoms and molecules of air. The ionosphere limits ground-based radio astronomy to frequencies above about 10 MHz. Karl Jansky made his historic observations at 20 MHz, close to the limit. Had he chosen a somewhat lower frequency he would not have been able to detect the Milky Way. Satellites in high orbits (above 6000 kilometres) can extend this limit down to 300 kHz and at 100 000 kilometres from earth frequencies as low as 25 kHz (a wavelength of about 12 kilometres) could be detected from the cosmic landscape. However, the interplanetary gas reaching out from the sun past the earth prevents lower frequencies than this arriving at the vicinity of the earth.

At the very-high-frequency (very-short-wavelength) end of the radio band, the molecules of gas in the earth's atmosphere start to absorb the waves. . . .

Van Allen Belts

As we leave earth's atmosphere, we see a faint glow of free–free radiation from the ionosphere. Further out, at about 30,000 to

100,000 kilometres from earth's surface we see the Van Allen 'radiation belts', doughnut-shaped zones of charged particles girdling the earth's equator. The Van Allen belts are formed by electrons and protons arriving from the sun at very high velocities—close to the speed of light—and becoming trapped by the earth's magnetic field. When a charged particle runs into a magnetic field, it experiences a force that is perpendicular both to its direction of motion and to the direction of the magnetic field. The particle is therefore forced into a circular or spiralling motion around the magnetic field. Many of the charged particles arriving from the sun find themselves channelled into a zone girdling the earth's equator. They spiral backwards and forwards along the magnetic field lines but they are trapped. Now, when the particles move close to the speed of light an effect of the theory of relativity comes into play and they radiate strongly as they spiral along the magnetic field. This radiation is called synchrotron radiation and we encountered it in our visible voyage in the jet in the nucleus of M87. Particles moving at speeds close to that of light are called 'relativistic' particles. In the radiation belts the relativistic particles spiral to and fro frenetically, radiating synchrotron radio emission. The discovery in 1958 of the radiation belts by the group led by James Van Allen of the University of Iowa, using the *Explorer I* satellite, marked the birth of a whole new area of space science, that of magnetospheric physics. This is the study of the region of space controlled and shielded by earth's magnetic field, the 'magnetosphere'.

Buffeted by Solar Wind

We see that a stream of new relativistic particles rushes towards the earth from the sun in a great wind, some to be trapped in the radiation belts, most to be deflected round the earth by its magnetic field, a few to travel on towards the unprotected polar regions to dissipate their energy in the magnificent spectacle of the aurorae. These wonderful displays of radiation generated by particles moving close to the speed of light give us a vivid feel for the way the solar 'wind' buffets the earth. This wind of gas and fast particles was discovered in the same year as the Van Allen belts and has completely changed our ideas about the space between the sun and the earth. Gas is continuously accelerated from the surface of the sun and fills interplanetary space with an ever-expanding and accelerating wind. The surface of the moon is directly blasted by this wind, but the earth's magnetic field creates a cavity round which the wind is swept, called the 'magnetopause'.

Perhaps we are travelling when the sun is disturbed by spots and flares on its surface. The solar wind becomes a gale, the radiation belts buckle to and fro, the ionosphere is buffeted about. Compass needles around the earth go haywire as a magnetic storm rages. Radio communications are wiped out and the radio band is flooded with bursts of radio emission from the direction of the sun. These radio bursts are caused by electrons moving close to the speed of light, accelerated in the flare event.

Yet when the storm is passed the sun has a pallid and bloated appearance in our band. As we travel towards the sun we find that its radio emission comes from far beyond the surface defined by visible light, from a zone known as the corona, after its crown-like appearance in visible light during eclipses. In visible light the disc of the sun is evenly bright, but in radio light the sun is brighter towards its rim than its centre, and is brighter towards the equator than towards the poles. The sun is not specially dazzling to our radio eyes, and we can hardly pick out any of the other familiar stars at all.

Jupiter's Radio Broadcasts

As we travel on outwards, we notice that only Jupiter stands out amongst the solar system planets. Jupiter's surprisingly strong radio emission was discovered accidentally by two US astronomers, B.F. Burke and K.L. Franklin, in 1955. They found inermittent bursts at low radio frequencies (around 20 MHz) which may be generated by lightning flashes in Jupiter's atmosphere and ionosphere. Curiously the brightness of these bursts of radio emission is affected by the passage of Jupiter's satellite, Io. At higher radio frequencies (wavelengths of tens of centimetres) some of the heat from the cool disc of the planet is seen (most comes out at infrared wavelengths), but also strong synchrotron radiation from relativistic particles trapped in Jupiter's radiation belts. Jupiter's magnetic field is ten times stronger than the earth's, its magnetosphere is a hundred times larger, and its radiation belts far more impressive than the Van Allen belts round earth.

The other planets can be seen only dimly in the radio band. Their temperatures range from warm to cool and most of their heat, or thermal energy, is radiated in the infrared, so we shall see them more clearly on our infrared voyage. However part of their thermal radiation is emitted as radio waves and so we do see a faint glow from them.

As we look round the radio sky, the stars have vanished and the brightest radio sources are in directions we had hardly noticed at all in the visible band.

The Milky Way

Dominating the sky is the radio emission from the Milky Way, which spreads over almost the whole sky at the lower radio frequencies. As we trace the radio photons back to their source we find they are being radiated by relativistic electrons spiralling in the Galaxy's large-scale magnetic field. These same electrons are among those atomic particles striking the earth which earlier in this century were given the name 'cosmic ray'. Cosmic rays consist of electrons, protons, and ions of other atoms (the ion of hydrogen is the proton) all moving relativistically. The most energetic of these have speeds within one part in a million million of the speed of light. Where have these cosmic messengers come from? Most come from the sun as part of the solar wind. The gaseous part of the solar wind streams past the magnetopause, the cavity in the wind created by the earth's magnetic field, while the lower energy relativistic particles are trapped in the radiation belts and the higher energy particles stream onto the earth as cosmic rays. The remainder of the cosmic rays come from much further away and permeate the whole of our Galaxy. The Galactic cosmic-ray electrons are the particles responsible for the radio emission from the Milky Way. These probably originate in supernovae remnants. . . . A few of the highest energy cosmic rays may come from outside the Galaxy, from explosions on a galactic scale we have not yet encountered in the cosmic landscape.

The Making of Stars

When we turn our attention in more detail towards the plane of the Milky Way, we find bright extended sources strung out along the Galactic plane like beads on a necklace. They stand out as very brilliant patches against the Milky Way at high radio frequencies, but at low radio frequencies they appear as dark absorbing patches. As we travel towards them we realize that many of them are the clouds of hot gas, heated by young, massive blue stars, that we noticed on our visible voyage. Both the radio and visible radiations are free–free emission from the ionized gas. These blue stars are actually radiating the bulk of their energy in the ultraviolet, as we shall see on our next voyage, and it is these energetic ultraviolet photons from the star which strip the electrons off the atoms of gas and ionize it. As we trace these clouds round the Galaxy we are witnessing the locations of recent star formation, for the associated stars do not live for more than ten million years or so. The clouds are not spread evenly through the disc of our Galaxy. Instead they are concentrated in a spiral pattern which can be followed right round the disc. If you trace out the Milky Way on the sky from Cassiopeia through Taurus to near Orion, you are looking at the spiral arm nearest to earth, only a few hundred light-years away.

If we make our voyage in a very particular radio wavelength, 21 centimetres, we suddenly see a different and much more intense sky. The reason is that atomic hydrogen has a characteristic wavelength here. Spiral arms stand out very strongly as long chains of clouds of cold gas. We realize that these are the reservoir of material for new stars and the spiral arms are the places where the clouds contract to form stars.

Discovering Proof for the Big Bang Theory

By Marcus Chown

Edwin Hubble's 1929 discovery that the universe is expanding had profound implications. To many scientists it suggested that at some time in the distant past, the universe must have been infinitely small. This inference led to a fierce debate between proponents of the theory that the universe burst out into existence at some point in the remote past and those who believed that the universe has always existed. The first viewpoint came to be known as the big bang theory, the second as the steady state theory. As often happens in modern astrophysics, these theories were proposed well before decisive proof for either was available.

In the following selection science writer and former radio astronomer Marcus Chown describes how evidence in favor of the big bang was discovered. In an interesting coincidence a pair of researchers at Bell Labs, Arno Penzias and Robert Wilson, who were frustrated by a persistent hiss in their horn-shaped antenna, learned about a group of astronomers at Princeton who were searching for just such a hiss in the skies. After conferring with the Princeton astronomers, the Bell Labs pair decided that they had stumbled onto the microwave afterglow of the big bang. Penzias and Wilson published their findings in 1965 and later received the Nobel Prize in physics for their achievement. However, the discovery was marred by controversy. The Princeton team, which published a theoretical paper explaining Penzias and Wilson's discovery, failed to credit the cosmologists who had originally predicted the existence of the big bang's leftover background radiation: cosmologist George Gamow and two of his doctoral students, Ralph Alpher and Robert Herman.

Marcus Chown, *Afterglow of Creation: From the Fireball to the Discovery of Cosmic Ripples.* Sausalito, CA: University Science Books, 1996. Copyright © 1996 by University Science Books. Reproduced by permission.

The oversight left Gamow and his students permanently embittered.

Marcus Chown is an award-winning science writer and broadcaster. Formerly a radio astronomer at the California Institute of Technology in Pasadena, he now writes on cosmology for *New Scientist* magazine.

I t is a fair claim that one of the greatest scientific discoveries of the twentieth century was made by telephone. In fact, not by one telephone call but by two.

In April 1965, [Bell Labs researcher] Arno Penzias phoned Bernie Burke, a prominent American radio astronomer at the Carnegie Institution's Department of Terrestrial Magnetism in Washington DC. Penzias's call was prompted not by the problem with the 20-foot antenna but by another matter altogether, and he would never have mentioned the irritating static had Burke not asked him in passing how the experiment on Crawford Hill was going.[1] Immediately, Penzias launched into a long complaint about the irritating signal that would not go away and about how frustrating it was trying to track down its source.

Burke sat up. One of his colleagues, Ken Turner, had told him about a search that was under way for just such a signal at Princeton. Could that be what Penzias and Wilson had picked up?

He tried to recall what Turner had told him. Turner had been to a talk the previous month given by [astronomer] Jim Peebles, a friend from his days as a graduate student at Princeton (Turner's supervisor had been none other than [Princeton astronomy professor] Bob Dicke). The talk Peebles had given was at a meeting of the American Physical Society held at Columbia University. As far as Burke could remember from what Turner had told him, it was about fireball radiation being an unavoidable consequence of a hot Big Bang. Peebles had argued that if the Universe's helium was indeed produced in the Big Bang, then today's Universe should be filled with microwaves with a temperature of less than 10 degrees above absolute zero. This tepid afterglow of creation was detectable with current technology. In fact, Dicke's group at Princeton had already embarked on a search for it.

1. Since 1963 Penzias and his colleague, Robert Wilson, had been operating a 20-foot horn-shaped antenna at Bell Labs' research station on Crawford Hill, in Holmdel, New Jersey, to pick up satellite signals. For several years they had been trying to eliminate the static, which, in the view of many, turned out to be an echo of the big bang.

Burke immediately alerted Penzias to the possibility that the anomalous signal might be the leftover glimmer of the Big Bang. It was music to Penzias's ears. By now he was desperate to find an explanation—any explanation—for the 3.5 degree excess temperature. He got on the phone to Dicke.

Beaten to the Punch

When the phone rang in Dicke's office at Princeton, Dicke had company. Seated in a circle around his desk, sipping cups of coffee and eating sandwiches, were his three disciples—[David] Wilkinson, [Peter] Roll and Peebles. "Every week we used to have these brown-bag lunches to chat about how our experiment was going and talk about what we ought to be doing next," says Wilkinson. "Arno's call came during one of those gatherings."

Dicke's telephone conversation was rather one-sided. He mostly listened, now and then nodding and repeating phrases familiar to the others in the office. Wilkinson's ears pricked up the moment he heard Dicke mutter the words "horn antenna."

Peebles remembers the conversation vividly. "I seem to recall it involved such mysterious things as pigeon droppings," he says.

Nobody in the Princeton group knew Arno Penzias or Robert Wilson [of Bell Labs] but the team was well aware of the 20-foot antenna Bell Labs had built out at Holmdel [New Jersey] for the Echo project. Roll and Wilkinson had learned about it while scouring the microwave journals before starting on their experiment. "It was abundantly clear to us that Bell Labs had the best antenna around," says Wilkinson. . . .

On the telephone, Dicke continued to repeat familiar microwave phrases. Then suddenly he said, "cold load."[2]

"As soon as we heard those words, we knew the game was up," says Wilkinson.

Moments later Dicke hung up the phone. He turned to Peebles, Roll and Wilkinson. "Well boys," he said, "we've been scooped!". . . .

Wilkinson and the others pored over Penzias and Wilson's data—wiggly red lines on chart recorders. By now, they were satisfied with what they had seen. "Penzias and Wilson were looking at a wavelength where there shouldn't have been any signal at

2. To ensure that they were picking up a signal from space, the researchers compared it with radiation given off by liquid helium chilled to 4.2 degrees Kelvin, called a "cold load."

all so we were convinced they must be seeing the primeval fireball," says Wilkinson.

The effect they had measured was small—no more than a few degrees. Any other instrument in the world would have missed it, but the Holmdel antenna was uniquely suited for distinguishing a weak background signal from other, much stronger sources. There on the chart recorder was a cryptic message from the very beginning of time.

Trace of the Fireball

If they were right, it was the most important discovery in cosmology since Edwin Hubble had found that the Universe was expanding back in 1929. Permeating every pore of the Universe was a tepid radiation, the "afterglow" of the titanic fireball in which the Universe was born. Before the Holmdel antenna had intercepted it, the radiation had been streaming across empty space for an incredible 15 billion years.[3] Penzias and Wilson had stumbled on the oldest "fossil" in creation, carrying with it an imprint of the Universe as it was soon after the creation event itself.

The temperature of the background radiation was the temperature of the early Universe, greatly reduced by the enormous expansion it had undergone since.

When the radiation broke free of matter, the Universe was at a temperature of about 3,000 degrees. But while it had been flying to us across space, the Universe had expanded a thousand times in size, diluting the temperature of the radiation by exactly the same amount so that today it appeared to be only about 3 degrees above absolute zero.

The temperature of about 3 degrees above absolute zero is the temperature of the Universe. Although the stars are very hot and very numerous, when their temperatures are averaged over all of space, their contribution to the temperature of the Universe is completely negligible compared with the fireball radiation.

The cosmic background radiation came from the time when it first became cool enough for atoms to form. At this instant, about 300,000 years after the Big Bang, the rapidly cooling fireball suddenly became transparent to light. Photons which had bounced from particle to particle in the fog of the fireball were suddenly

3. The estimate of the age of the universe has since been revised to 13.7 billion years.

able to move freely. And they have been doing so ever since, gradually losing energy as the Universe has grown in size.

It may seem peculiar that the cosmic background radiation is arriving at the Earth only today, 15 billion years after the Big Bang. After all, in a sense we were in the Big Bang (or at least, the particles of matter that would one day condense to form the Earth were in the Big Bang) and the fireball radiation was all around us. Surely it should have already passed us by now?

Long Journey

Well, radiation which in the Big Bang was emitted by matter in our immediate neighborhood has already passed us. Forgetting for a moment that the Universe has expanded a lot since the Big Bang, it is true to say that radiation emitted 15 billion light years from us is just arriving at the Earth today. On the other hand, radiation that was emitted 10 billion light years away would have arrived 10 billion years after the Big Bang—or just as the Sun and Earth were forming 5 billion years ago.

The expansion of the Universe complicates matters a little because when those photons of the Big Bang radiation arriving at the Earth today broke free of matter, the Universe was only a thousandth of its present size. The photons have therefore taken 15 billion years to cross a gap that was originally only 15 million light years wide. It is as though you were trying to sprint 100 meters on a track that has grown a thousand times longer while you are running.

The detection of the cosmic background radiation by Penzias and Wilson meant that the Big Bang theory was triumphant. If Martin Ryle's work at Cambridge on radio galaxies had sent the Steady State theory reeling, the discovery of the afterglow of creation dealt it a knock-out blow.

For the second time in its history, scientists at Bell Labs in Holmdel had made a great scientific discovery serendipitously. Back in 1931, a 26-year-old Bell Labs physicist named Karl Jansky, who had been investigating possible sources of radio interference, detected a weak static that seemed to be coming from the Milky Way, and thus invented the science of radio astronomy. . . .

Penzias and Wilson were both slow to accept the cosmological origin of their mysterious signal. "They'd spent so long focusing on all the mundane explanations—like pigeon droppings," says

Peebles, "that I think it took them a while to realize just how great a discovery they had really made."

In fact, it was at least a year before the two astronomers would accept that their anomalous signal came from the Big Bang. "We had made a measurement which we thought would hold up," says Wilson. "But we weren't so sure that the cosmology would."

Wilson had another reason for dragging his feet. "I'd rather liked the Steady State theory," he says. Inadvertently, he had helped to destroy it.

But though Penzias and Wilson were a bit dubious about the Big Bang idea, both were very pleased to finally have an explanation for the problem that had been troubling them for so long. "When we came along, they were at a complete loss for any other explanations," says Peebles. "They were feeling driven against a wall."

"They desperately wanted to use the antenna to do some radio astronomy," says Wilkinson.

This is certainly illustrated by Penzias's immediate reaction to the Princeton explanation. According to Peebles, in one of their early telephone conversations, Penzias said: "Well, that's a big relief. We understand this thing at last. Now we can forget it and go and do some real science!" But rarely had there been a scientific result that was less likely to be forgotten!

The parallels with the century's other great cosmological discovery were striking. Both the expansion of the Universe and the fireball radiation had been found by scientists who were completely unaware that predictions of the phenomenon had been made many years before in the scientific literature. Which makes you wonder whether scientists ever read the scientific literature at all—or remember what they read!

Putting It in Print

The Princeton and Bell Labs groups decided to announce the discovery in two scientific papers, published side by side, in *Astrophysical Journal Letters*.

Two weeks before the papers were due to appear in print, Wilson finally began to realize how important a discovery he and Penzias had made. The phone rang out at Crawford Hill, and on the other end was Walter Sullivan, the science reporter of *The New York Times*.

Sullivan had been on the trail of another story entirely when he had happened to call the offices of *Astrophysical Journal.* "For some unknown reason they leaked our paper to him," says Wilson. Sullivan grilled Penzias about the work with the 20-foot antenna.

At the time of the phone call, Wilson's father was visiting him from Texas. An habitual early riser, the next day he got up well before his son to walk down to the local drugstore. When he came back, he had a copy of *The New York Times*. He thrust it in the face of his bleary-eyed son. There on the front page was a picture of the 20-foot horn with a description of the *Astrophysical Journal* paper. "For the first time, I really got the impression the world was taking this thing seriously," says Wilson.

No Mention of Predecessors

[Cosmologist] George Gamow, by now retired, read the story in *The New York Times*. To his dismay, he saw no mention of his name, nor those of Ralph Alpher or Robert Herman. It is fair to say that he awaited the publication of the scientific papers with intense interest.

The papers duly came out. The title of Penzias and Wilson's paper gave nothing away: "A Measurement of Excess Antenna Temperature at 4800 Megacycles per Second." Rarely can such an important scientific discovery have been disguised so well!

In the paper, the two Bell Labs astronomers wrote: "Measurements of the effective zenith noise temperature of the 20-foot horn-reflector antenna at the Crawford Hill Laboratory, Holmdel, New Jersey, at 4080 megacycles per second have yielded a value of about 3.5 degrees higher than expected."

And that was basically all Penzias and Wilson said. Nowhere in their brief paper did they mention that the radiation they had picked up might have come straight from a hot Big Bang. They merely noted: "A possible explanation for the observed excess noise temperature is the one by Dicke, Peebles, Roll and Wilkinson in the companion letter in this issue."

"I think they were rather overcautious," says Wilkinson.

"Their paper was written in such a way that it could have been almost anything they'd found," says Dicke.

"In contrast, our group really went out on a limb," says Wilkinson. "In our paper, we were interpreting a single microwave measurement as proof of the existence of the Big Bang radiation."

"In fact, Penzias and Wilson weren't even going to write a paper at all until we told them we were writing one," says Dicke.

Wilson says the reason he and Penzias did not write about the Big Bang theory of the origin of the background radiation was because they were not involved in that work. "We also thought that our measurement was independent of the theory and might outlive it," he says.

"We were pleased that the mysterious noise appearing in our antenna had an explanation of any kind, especially one with such cosmological implications. Our mood, however, remained one of cautious optimism for some time."

The Gamow Controversy

The moment the two scientific papers were published, Gamow made a beeline for his library. He raced through the two papers, becoming increasingly angry. Nowhere was there a mention of his ground-breaking work in the 1940s. Gamow, [and his proteges Ralph] Alpher and [Robert] Herman had not only published the results of their hot Big Bang calculations in a series of technical articles in *The Physical Review* but they had written numerous popular accounts of their work as well. For instance, in 1952 Gamow published a book for lay readers called *The Creation of the Universe* in which he talked about the cooking of helium in a hot Big Bang and how this was connected to the temperature of the Universe. Four years later, Gamow aired his ideas in an article in the popular magazine *Scientific American.*

But all these accounts were missed entirely by Dicke's team at Princeton. "We absolutely didn't know about Gamow's work," says Wilkinson. "When Jim Peebles and I were searching through the scientific literature to see what had already been done, we read only the microwave journals so we never saw any of Gamow's stuff."

One of the problems was that before Penzias and Wilson's discovery of the cosmic background radiation, cosmology was not really a distinct field. "There was no cosmology literature," says Wilkinson. "The scientific papers that were published—and there were not many—were published all over the place. Even today, thirty years later, I'm still finding papers on the cosmic background radiation that I never knew existed."

But though it is easy to understand how Wilkinson and Peebles missed Gamow's work, it is harder to explain how Dicke could have missed it. Several years earlier [early 1960s] he had attended a talk Gamow had given at Princeton about making elements in a hot Big Bang. "Gamow spoke about a Universe in which you start with a mass of cold neutrons which suddenly explode in a Big Bang," he says. "But that's all I can recall about what he said."

And the connection between Dicke and Gamow does not end here. It turns out that the very same issue of *The Physical Review* that contained George Gamow's first 1940s paper on the hot Big Bang also contained a paper by Dicke. That might not seem too much of a coincidence but, in a throwaway remark in his own paper, Dicke actually made a comment about the possibility of a microwave background radiation in the Universe.

Memory Failure

As part of his wartime radar work, Dicke and his colleagues had gone to Florida to measure the radio waves coming from water vapor in the moist atmosphere. As an aside, he had wondered whether the sky might be glowing uniformly with microwaves. If such a uniform glow existed, it would have to be coming from the Universe as a whole since nearby sources, such as a planet or the Milky Way, would fill only a small part of the sky.

Dicke concluded that there was no such sky-glow that he could measure. In fact, he put it more precisely in his paper in *The Physical Review*, stating that the temperature of any "radiation coming from cosmic matter" had to be less than twenty degrees above absolute zero.[4]

Dicke had thus attempted the first ever measurement of the Universe's radiation background. But ironically he had forgotten all about it, and so, too, had everyone else. "Jim stumbled on it only when we were reading through the microwave literature," says Wilkinson. In the cosmic background field not only did people often overlook each other's work, they sometimes even overlooked their own!

But such forgetfulness was hardly likely to console Gamow, Alpher and Herman. The irony was that the last thing anyone wanted to do was upset Gamow. He was an idol to the young radio astronomers at Princeton and Bell Labs.

---

4. The technique Dicke used and the receivers available in the 1940s were not capable of detecting a uniform background as cold as three degrees above absolute zero.

"Gamow was one of my heroes," says Wilkinson. "I read all of his popular books in high school. He was probably the reason I got into science in the first place." Robert Wilson had also been turned onto science by reading Gamow's popular books.

All of them realized that Gamow was one of the most intuitive and inventive physicists of the twentieth century. "He had the ability to ferret out the essential elements of the most complicated physics," says Peebles. "It was that ability he used to effect when tackling the problem of the Big Bang and the fireball radiation."

Unforgiven

Peebles and the rest felt guilty that they had not given due credit to Gamow's group. "We simply did not do our homework," he says. "We should have gone through the literature and got every possible reference to this thing. In fact, it was a couple of years before we did that." This failure to right the wrong immediately ensured that Gamow, Alpher and Herman would remain bitter about the way they had been treated.

"I tried to do all I could to bring Gamow into the whole story as much as possible," says Wilkinson. Soon after the momentous events of spring 1965, he and Peebles decided to write an article about the discovery for the magazine *Physics Today*. Before putting pen to paper, they went back and read the papers of Gamow, Alpher and Herman. But the article never got past the rough draft stage. "Alpher and Herman took issue with our version," says Wilkinson. "They wrote us a rather strong letter. So in the end we withdrew the article and never published it."

Perhaps, if someone in the Princeton team had actually telephoned Gamow at the outset, and asked him just what his group had done and when, then all the misunderstandings would have been avoided.

Arno Penzias tried his best to smooth things over with Gamow but feelings were simply running too high. "I don't think Gamow ever really forgave Dicke and his group," says Wilson. "As for us, I don't know exactly how he felt."

Gamow remained bitter until his death, in 1968, just three years after the definitive proof of the hot Big Bang he had championed. "Alpher and Herman never got over it completely either," says Wilson.

CHAPTER 3

Astronomy in the Age of Space Exploration

Landing on the Moon

By Buzz Aldrin

The 1969 landing of astronauts on the moon is widely considered the greatest achievement in manned space exploration to date. Coming just eight years after President John F. Kennedy challenged America to send a crew to the moon within a decade and return them safely to Earth, the *Apollo 11* mission amazed the world.

In the following selection Edwin "Buzz" Aldrin, who along with Commander Neil A. Armstrong piloted the lunar module *Eagle* to a safe landing on the moon, recounts the landing. It was a tense affair. Just moments into the descent, Aldrin recalls, an alarm went off, putting the landing in jeopardy. Trusting in the guidance of Mission Control in Houston, the astronauts proceeded. Armstrong, seeing rubble below them, moved the *Eagle* horizontally in search of a safe landing spot. Aldrin struggled to suppress a sense of panic as alarms continued to sound and the lunar module's landing fuel ran low. (A separate tank held fuel for ascent.) The pair made a good landing, however, with just seconds of landing fuel left in their craft.

The astronauts went on to complete a historic mission. Among its contributions to astronomy were numerous unique images photographed on the surface of the moon and the transference of nearly fifty pounds of lunar rocks to Earth for further study. Aldrin became the second man, after Armstrong, to walk on the lunar surface. Their voyage was followed by several more, but after *Apollo 17* landed a pair of astronauts on the moon in late 1972, the program was halted. That marked the end of human exploration of the moon. However, as of 2005 the National Aeronautics and Space Administration is planning a new round of manned lunar missions with the goal of eventually sending astronauts to Mars. Lieutenant Colonel Buzz Aldrin, a former test pilot in the U.S. Air Force, retired as an astronaut and went on to become a leading advocate for space exploration.

Buzz Aldrin, "The Eagle Has Landed," *Men from Earth,* by Buzz Aldrin and Malcolm McConnell. New York: Bantam Books, 1989. Copyright © 1989 by Research and Engineering Consultants, Inc., and Malcolm McConnell. Reproduced by permission of the author.

L unar Module Eagle: 20 July 1969
   As the spacecraft flew backward, Neil [Armstrong] and I
   watched the green digits blink on the small computer screen.
We arched westward along the equator of the Moon. Through the
triangular windows of the lunar module, we could see a chaotic
map of endless gray craters.

At this point, there was no sensation of falling. We glided, our
faces parallel to the surface. We were near "perilune"—the lowest point on our coasting orbit—ten miles over the silent craters.
Soon the computer would initiate the final twelve minutes of the
landing attempt, powered descent.

"Eagle, Houston," the capcom [chief mission communicator],
Charlie Duke, called from Mission Control, a quarter million
miles away. "If you read, you're go for powered descent."

Charlie's voice was barely distinguishable in my headset. Since
coming around the limb of the Moon from the far side, our voice
and data link with Houston had been shaky. Mike Collins, orbiting fifty miles above us in the command module Columbia, heard
Houston clearly.

"Eagle, this is Columbia," Mike called, his voice calm. "They
just gave you a go for powered descent."

Descent Begins

The green digits flashed, announcing PDI [powered descent ini-
tiated]. Beneath our feet, hypergolic propellants sprayed into the
combustion chamber, igniting on contact and spewing a soundless
plume of orange flame into the sunlit vacuum of space. The lunar
module slowed as the computer throttled up the engine from a 10
percent thrust to full power. The pocked moonscape rolled closer
in my window. Four minutes into the burn, I began to feel the
weight in my arms caused by our first prolonged sag of decelera-
tion, a subtle imitation of gravity. My pressure boots flexed and
my limbs settled inside my suit.

"Eagle, Houston. You are go," Charlie Duke announced, his
voice clearer now through the hissing static. "Take it all at four
minutes. Roger, you are go to continue powered descent. . . ."

The horizon moved across my window and settled at the bottom as the LM [lunar module] completed its slow rotation on its
vertical axis. A gradual pitch-over would bring the lander to its final descent posture, with our feet toward the lunar surface. Now

we could see Earth, a partial disk of blue, white, and brown hanging far above the serrated horizon of the Moon.

While the engine fired, the pitch-over continued and Earth slipped past the top of my window. My instrument panel was now alive with winking data. We passed through 35,000 feet. For the first time we felt the rapid descent.

Alarm Sounds

At that moment, an alarm we weren't prepared for flashed on the top row of our computer data screen. "Twelve-oh-two," I called, unable to control the tension in my voice. "Twelve-oh-two."

Neil and I exchanged quizzical, troubled looks. We were descending through 33,000 feet, and our primary computer had just signaled difficulty coping with the cascade of data coming in from the landing radar. The data screen went blank. All we could do was wait for the experts in Mission Control to decipher this alarm, but they were more than two light seconds away, which meant their reaction to the telemetry data on the alarm would not be immediate. To me the alarm was ominous. Either the programming was incapable of managing the landing, or there was a hardware problem. In either case, the potential for catastrophe was obvious because our eventual ascent from the surface in twenty-four hours and our rendezvous with Columbia would place even greater demands on that computer. Still, we coasted toward the Moon.

"Give us the reading on the twelve-oh-two program alarm," Neil called, voicing the tension we shared.

We both eyed the large red ABORT STAGE button on the panel. Hitting it would instantly blast the LM's bulbous upper stage back toward Columbia, ending man's first attempt to reach the moon.

"Roger," Charlie called, the strain in his voice obvious. "We've got . . . we're go on that alarm." Charlie's transmission shredded with static again.

I felt the immense gulf separating the brightly lit control room in Houston and the dim cabin of the LM. Neil nodded to me through his helmet, his eyes somber in the glow of the panel. Charlie's terse "go" meant the guidance officers at Mission Control had judged the problem an acceptable risk. The time skip in our communications made a discussion of the situation impossible at this critical point. We simply had to trust Houston. Another alarm went off on the data screen. I fought the urge to shout a warning to Neil. Even after years as a fighter pilot and an astronaut, I felt the first hot edge of panic. I knew there had to be something seriously wrong with our guidance computer, and yet we were still descending. Again Mission Control reassured us with a "go" call. They had dismissed the alarm as noncritical, but they couldn't take the time to explain why. We dropped through 20,000 feet, eating up altitude at 150 feet a second. As the LM continued tilting forward, our triangular windows filled with craters and humped ridges. During this phase we had to be sure we were heading toward a safe landing spot, but all we saw were gray craters and boulder fields. Again the alarms flashed, banishing data from our screen.

Tense Moments

"Twelve alarm," Neil called. "Twelve-oh-one." "Roger," Charlie acknowledged. "Twelve-oh-one alarm."

I licked my dry lips. This was a time for discipline. But the tension had me rigid inside my suit. We had to trust Mission Control.

"We're go," Charlie added. "Hang tight, we're go. Two thousand feet. . . ."

The pitch-over maneuver continued, giving us an excellent view of the rills and craters below us. Neil hunched over his hand controller knob, ready to take command if the automatic landing system led us into danger. I began a continuous series of readouts, giving both Neil and Mission Control a detailed description of our final descent.

At five hundred feet, Neil was not satisfied with the landing zone. He took over manual control from the computer, slowing our descent from twenty feet per second to only nine, and then at three hundred feet, to a descent of only three and a half feet per second. The LM hovered above its cone of flame. Neil did not like what he saw below. He delicately stroked the hand controller like a careful motorist fine-tuning his cruise control. We scooted horizontally across a field of rubbly boulders. Two hundred feet and our hover slid toward a faster descent rate.

"Eleven forward, coming down nicely," I called, my eyes scanning the instruments. "Two hundred feet, four and a half down. Five and a half down. One sixty. . . . Quantity light."

The amber low-fuel quantity light blinked on the master

caution-and-warning panel near my face. The lumpy horizon of the Moon hung at eye level. We were less than 200 feet from landing on the Moon, but Neil again slowed the descent.

Simultaneously, Charlie's voice warned, "Sixty seconds."

The ascent engine fuel tanks were full, but completely separate from the big descent engine. We had a maximum of sixty seconds of fuel remaining in the descent stage before we had to land or abort. But Neil had slowed to a hover again as he searched the ground below.

"Down two and a half," I called. "Forward. Forward. Good. Forty feet. Down two and a half. Picking up some dust. Thirty feet. . . ."

Thirty feet below the LM's gangly legs, dust that had lain undisturbed for a billion years blasted sideways in the plume of our engine.

"Thirty seconds," Charlie announced solemnly, but still Neil hovered.

Touching Down

The descent engine roared silently, devouring the last of its fuel supply. Again, I eyed the ABORT STAGE button. "Drifting right," I called, watching the shadow of a footpad probe straining to touch the surface. "Contact light." We settled silently onto the Moon, with perhaps twenty seconds of fuel remaining to fuel the descent stage. Immediately, I began preparing the LM for a sudden abort ascent in the event the landing had damaged the Eagle or the lunar surface was not strong enough to support our weight.

"Okay, engine stop," I told Neil, reciting from the checklist. . . .

"Got it," Neil answered quietly, disengaging his hand control system.

"Mode controls, both auto," I continued. "Descent engine command override off. Engine arm off.". . .

"We copy you down, Eagle," Charlie Duke interrupted from Houston.

From the window before my face I looked out over the alien rock and shadow of the Moon. I breathed inside my helmet, totally absorbed. A mile away, the horizon curved into blackness.

"Houston," Neil called, "Tranquillity Base here. The Eagle has landed."

Uncovering the Fabric of the Universe: Dark Matter and Dark Energy

By Michael S. Turner

In the following selection astronomer Michael S. Turner reviews some of the strange phenomena found in the universe. In 2003 the Wilkinson Microwave Anistrophy Probe (WMAP), a space-based observatory used to measure cosmic background radiation, furnished evidence for an astonishing hypothesis. Astronomers are now convinced that ordinary matter and energy—the material that makes up stars, planets, and comets—comprise less than 1 percent of the content of the universe. According to Turner, dark matter, the material whose gravitational tug holds galaxies and clusters of galaxies together, comprises about a third of the universe's mass. No one knows just what dark matter is, but its existence has been confirmed through observation of its effects. Turner explains that astronomers are struggling to identify the nature of dark matter. It may be composed of neutrinos, which are extremely hard to detect, or even more exotic particles, such as neutralinos or axions, which have never been observed. As bizarre as dark matter may be, an even greater mystery has cropped up: dark energy. Turner says that dark energy accounts for as much as 66 percent of the content of the universe, and yet almost nothing is known about it, apart from the fact that it is driving galaxies apart at a growing rate. Michael S. Turner is chairman of the Department of Astronomy and Astrophysics at the University of Chicago.

Michael S. Turner, "Absurd Universe," *Astronomy,* vol. 31, November 2003, p. 44. Copyright © 2003 by Astronomy Magazine, Kalmbach Publishing Co. Reproduced by permission.

Place $2 worth of pennies in a pile. Select only one penny, preferably a nice, shiny one, and set it apart. Cover the pile of 199 pennies with a cloth. What you've created is a crude model of the universe, where what we see with our eyes is only a small part of what our brain knows to exist.

Sound absurd? On February 11, 2003, the first results from the Wilkinson Microwave Anisotropy Probe (WMAP), an orbiting observatory, showed us what we couldn't see. WMAP, conceived in 1995 and launched in June 2001, was a follow-up to NASA's Cosmic Microwave Background Explorer (COBE), which, in 1992, discovered tiny (0.001 percent) variations in the intensity of cosmic microwave radiation across the sky. These variations provided a map of sorts to what the universe was like some 400,000 years after its beginning. Using this map as a starting point, astronomers predicted the makeup of the universe. What they needed, however, was a better instrument to confirm the prediction.

WMAP unveiled—with remarkable resolution—a seemingly absurd universe comprised of 0.5 percent visible matter (stars, dust, and gas), roughly 33 percent dark matter that holds the universe together, and roughly 66 percent mysterious dark energy that is accelerating the expansion of the universe.

With the picture in clear view, scientists have developed a theory: The tiny patterns in the millionth-of-a-degree temperature fluctuations were the seeds of what later became galaxy clusters and other large-scale structures we observe today.

Dark Matter Rules

A few decades ago, the idea that the visible universe made up only 0.5 percent of all there is would have sounded ridiculous to most astronomers. Hints of our universe's strangeness, however, were found more than 70 years ago. Fritz Zwicky, the eccentric Swiss-American astronomer based at Caltech, pointed out in the 1930s that the individual members of the Coma cluster of galaxies were moving too fast to be bound by the gravitational effects of their stars alone. Without the additional gravity of some unseen form of "dark" matter—perhaps even in the form of billions of small bodies like Jupiter—the galaxies within the Coma cluster would have long since dispersed. Today, astronomers can map the amount of dark matter contained in many galaxy clusters by studying how those clusters bend light from more distant objects.

Cosmic Background Radiation

Clusters were strange, and Zwicky even more—so it was easy to ignore the dark matter problem. However, a complication arose from another direction. In 1964, Bell Labs scientists Arno Penzias and Robert Wilson discovered the cosmic microwave background (CMB) radiation. The Big Bang theory predicts that the early universe was a very hot place. As it expands, however, the radiation within it expands into a larger and larger space and cools. Today the universe is filled with the Big Bang's remnant heat—the CMB. The CMB contributes only negligibly to the energy content of the universe, about 0.01 percent of the total, but it provides an exquisite record of the universe at a simpler time.

Because the universe is expanding, its radiation is becoming weaker as photons fill an ever-larger space and as each photon's energy is redshifted [shifted to a longer wavelength by stretching]. Looking forward, this means the energy of the CMB will become less important; but looking backward, it means that earlier than 10,000 years after the Big Bang, the radiant energy of the CMB was very important. In the early universe, matter was the bit player.

Vera Rubin, an astronomer at the Carnegie Institution, and others brought the dark matter problem much closer to home in the 1970s and 1980s. With the discovery that both stars and clouds of gas farthest from the center of the galaxy are moving too fast to be held by the gravity of just stars, Rubin and her colleagues showed how dark matter actually holds galaxies like our own together. Even at our position in the Milky Way, about half the gravity that keeps us in orbit is due to unseen dark matter.

The late David Schramm showed how the production of the lightest elements, especially deuterium, in the Big Bang could be used to measure the total amount of ordinary matter. The conclusion was shocking—no matter what form it was in, there simply was not enough ordinary matter (and thus gravity) to hold the universe together. WMAP, because of its greater precision, found a clue to this puzzle in the cosmic microwave background radiation—while Carl Sagan was right when he said we are made of star stuff, cosmologists now know that the universe is not.

Dark Matter Candidates

The infusion of particle physics into cosmology in the 1980s brought a revolutionary idea: Dark matter may exist as particles of a new form of matter. Three particle suspects have emerged: the neutrino (actually known to exist), the neutralino (100 times heavier than a proton), and the axion (one-trillionth the mass of an electron). For now, the neutralino and axion exist only in the minds of particle theorists and cosmologists.

The particle dark matter idea got a boost in 1998. The Super-Kamiokande detector in Japan, studying neutrinos produced in Earth's atmosphere by cosmic rays and in the interior of the Sun, showed that neutrinos have mass, albeit very little. Because so many neutrinos are left over from the Big Bang, if a single neutrino has even a tiny mass, all neutrinos combined become players in the cosmic mix. In the spring of 2002, astronomers at the Sudbury Neutrino Observatory in Ontario, Canada, studying neutrinos from the Sun, discovered more evidence that neutrinos have mass. And in the autumn of 2003, two experiments in Japan using human-made neutrinos further strengthened the case that neutrinos contribute somewhere between 0.1 and 5 percent of the total matter of the universe.

The story is not over yet. Neutrinos cannot account for all the dark matter; they appear to be just a cosmic spice. Particle physicists and cosmologists are pinning their hopes on the neutralino and axion, both referred to as cold dark matter. (This name comes from the assumption that dark matter particles were moving slowly when galaxies began to form.) Scientists are trying to produce neutralinos using particle accelerators. Ultra-sensitive instruments are trying to detect axions and neutralinos to see if, indeed, they exist.

Dark Energy

Just as many cosmologists and astrophysicists were getting used to the idea of particle dark matter and the fact that we are not made of the primary stuff of the universe, along came dark energy. Two groups of astronomers in the late 1990s, one led by Saul Perlmutter of Lawrence Berkeley Lab in California, and the other by Brian Schmidt of Mt. Stromlo Observatory in Australia, used distant supernovae as standard candles to measure the slowing of the expansion of the universe.

They found, however, that the expansion is not slowing down—it is speeding up—apparently due to the repulsive gravitational action of something that doesn't give off light. This "something" is referred to as dark energy because the way it reacts with the rest of the universe implies a form of energy. If the accelerated expansion continues for another 150 billion years, the sky will darken, with only a handful of galaxies remaining visible.

So here we are in our absurd universe. The Big Bang was a hot beginning, and its radiant energy controlled the universe's first 10,000 years. Stars illuminated the heavens, but they contributed only 0.5 percent of the total mass. The recipe for our universe today is crazier than even Zwicky could have imagined: one-third dark matter, two-thirds dark energy, a pinch of microwave energy, and at least three spatial dimensions. (String theorists insist there must be more.) Dark matter breaks down into 4 percent dark, ordinary matter (the bulk of it in the form of gas that is still too hot to form stars), 30 percent elementary particles, and the rest, simply referred to as cold dark matter, is still to be identified.

The past two decades have seen major advances in understanding how the universe is put together. If we are truly in a golden age of cosmological discovery—and I believe we are—big telescopes, space instruments, and accelerators will, over the next twenty years, help us make sense of this amazing universe in which we find ourselves.

Twin Rovers Investigate the Surface of Mars

By A.J.S. Rayl

In January 2004 two robotic explorers touched down on the surface of Mars. Launched by the National Aeronautics and Space Administration (NASA), they had been expected to function for about ninety days. Despite some problems, both rovers greatly exceeded their planned mission lives. In the following selection writer A.J.S. Rayl describes the many challenges NASA faced in bringing off the dual Mars explorer mission. History shows that two of every three missions to Mars fail, Rayl notes. Among the hurdles on this mission was a powerful solar flare that could have knocked out the rovers' electronics. *Spirit*, the first of the two rovers to land, experienced severe software problems for a time. In the other rover, *Opportunity*, mechanical problems cropped up. However, mission controllers overcame the difficulties. Both rovers, operating in different areas of Mars, have found evidence that water once flowed on Mars. Each has also found many novel rock and mineral formations. On April 5, 2005, as the solar-powered rovers continued to operate, NASA announced that it was extending their mission for as long as eighteen additional months. A.J.S. Rayl, a contributing editor for the *Scientist*, writes on health and science for a variety of publications.

As *Spirit* rang in her new year at Gusev Crater on Mars Monday [January 3, 2005], NASA [National Aeronautics and Space Administration] officials and mission team members celebrated the Mars Exploration Rovers' [MER] first anniversary at an event that featured a press conference, storytelling

A.J.S. Rayl, "NASA/JPL Team Celebrates Rovers' First Year on Mars as *Spirit* and *Opportunity* Return More Surprises," www.planetary.org, January 5, 2004.

session, and birthday party at the Jet Propulsion Laboratory (JPL).

"What a year it was," Charles Elachi, director of JPL and NASA Advanced Planning, mused before the hundreds of people who packed into Von Karman Auditorium to remember and 're-live' *Spirit*'s landing.

Spirit has begun her second year on Mars investigating rocks unlike any ever seen before on Mars. Her twin, *Opportunity*, which landed 20 days later on the other side of the planet, officially completes her first year January 23 [2005]. The rovers are funded through March, but there are no indications right now they'll be ready to stop then.

At an initial cost of about $820 million, the twin robot field geologists—which were designed and built and are being managed out of JPL—were guaranteed to last 90 days. Now, one year later, they're tolerating a few mechanical aches and pains—*Spirit* has to drive backward as much as forward because of a gimpy right front wheel—but all in all they are going strong. Each rover is now into her third mission extension and is returning small bounties of science data about our closest neighbor every week, at a combined cost of $3 million per month. . . .

Many Obstacles to Success

The MER mission has had all the ups and downs, all the drama and comedy, mystery, dilemmas, and conflict needed for any great epic that crescendos with a happy Hollywood ending—and it's been quite a ride.

Spirit and *Opportunity* and their team made it look easy to the world outside, but MER scientists and engineers, and NASA officials, told another story Monday during the press conference and storytelling event, both of which were broadcast live on NASA Television, as they 'relived' what was really going on behind the scenes and the stress they endured to what became the best of all possible endings. Those who were fortunate enough to have been there, journalists included, know well just how different the vibe in Von Karman Auditorium was only one year ago.

"The months leading up to launch were incredibly hectic." Rob Manning, the head of the Entry, Descent, and Landing team, set the stage in a pre-recorded video Elachi introduced at the beginning of the press conference. "We had so much to do and so little time to get it done . . . somehow we managed to get it done before launch. Even then the surprises weren't over. We had the worst solar flare reported in human history hit our vehicle and just days before landing on and Mars had a global dust storm that threatened both *Spirit* and *Opportunity*."

Then, one week before *Spirit* is to descend into the Martian atmosphere, the U.K.'s *Beagle 2* lander—dispatched by the European Space Agency's *Mars Express*—was lost. It reminded everyone of the cold, harsh reality that nearly 2 of every 3 missions that dare invade Martian space don't make it, and NASA officials prepared the global media for the worst from the "death planet." The MERs had the benefit of better testing and telemetry technology, and this team believed in its charges and they believed in themselves.

Even so, Manning at one point shared with the crowd gathered in the auditorium, that having to live in a two-faced world—outwardly projecting the dismal realities of travel to Mars while at the same time smilingly, encouraging the team internally to believe and soldier onward—was quite the challenge. It was all coming together though it seemed—and then the Red Planet presented *Spirit* with a different atmospheric profile than the team expected. Individual stress meters registered as high as humanly capable and the tension at JPL was as intense as it's ever been.

Someone apparently forgot to inform *Spirit* of that though. The rover soared through the alien atmosphere, releasing her parachute as commanded, just a bit early, and slowed as programmed from 12,000 miles per hour to zero in about 6 minutes, hitting the surface well-protected by airbags, and bouncing safely to an upright landing (and probably beaming bolt to bolt). Within a week, *Spirit* was preparing to rove, all systems seemed to be nominal and the team was breathing easier. Until—*Opportunity* was on final approach to Mars. Suddenly, *Spirit* clammed up and the team was "very concerned" that the communications breakdown "could be possibly very serious," then–project manager Pete Theisinger recalled.

Both Rovers Beat the Odds

"In fact, it was the morning of *Opportunity* landing that we found out that *Spirit* was going to survive the software problems that had been plaguing it the last week," Manning remembered. The best of Tinsel Town's scriptwriters couldn't produce drama any better than this.

An agonizing few hours later . . . ecstasy. A double-header. The

MER team had two rovers safe and sound and ready to go to work on the surface of Mars. It was, simply, spectacular.

"Then we got those first pictures from Meridiani [Planum, the region of Mars where *Opportunity* landed] and it was like nothing anybody had ever seen," remembered Steve Squyres, of Cornell University, principal investigator for the rovers' science payloads. "It showed unambiguously layered—*layered*—bedrock. The thing that we were hoping, praying that after hundreds of meters of driving, maybe we'd find—*maybe*—Bang! It's just a few meters in front of us."

If there was one moment, Squyres added later, from behind the dais in the auditorium, that was it—"not just because it meant we had two safe vehicles on Mars. But that moment when we opened our eyes and saw that layered bedrock 8 meters in front of us—man, that was *magic*."

Jim Erickson, rover project manager at JPL, agreed, and so, no doubt, would most everyone who was there at JPL watching exactly one year ago, for there was history being made in that magic.

"It's been almost a year since I sat right here and said that if things went well we were all about to embark on what was going to be arguably the coolest geology field trip in human history," reflected Squyres Monday. "I think the events have borne that out."

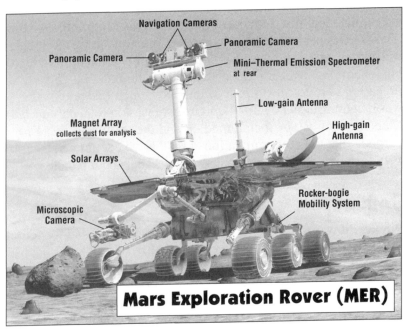

Mars Exploration Rover (MER)

Evidence of Water

Indeed. *Spirit* and *Opportunity* successfully achieved their prime directive last March, finding, as it turned out, that both Gusev Crater and Meridiani Planum boasted watery environments millennia ago. Not only had they hit pay dirt, they returned 'the gold' within the primary phase or first 90-days of their journey. Although the water at Gusev seems to have been underground, at Meridiani the evidence points to a salty sea that covered much of the surface at times. Suddenly, the notion of Martian life was real.

Opportunity uncovered in that bedrock before her clear evidence that rocks around her landing area had been deposited in flowing liquid water on the surface of Mars, and the announcement was a moment that will go down in the annals of science history. This is the first *in situ* evidence—data collected at Mars—that shows well beyond a reasonable doubt that the Red Planet once had an environment that could have supported life.

While space agency officials used the occasion to view MER in the context of the multiple successes last year, including *Cassini* and even *Genesis* which managed to return good research even though it crash-landed, there could be no underestimating the rovers' impact on space exploration last year. "What the exploration team had done in the last 12 months is nothing short of remarkable. The ongoing missions of the MERs are simply incredible," noted [NASA administrator Sean] O'Keefe, who is leaving his position as administrator soon for the post of chancellor of Louisiana State University. "When we finally got the signal that *Spirit* had safely landed in the Gusev Crater, the joyful celebration here did so much to lift the spirit of our citizens around this great country, as well as around the globe . . . it was a huge, huge event. Most importantly, it gave a boost to all of us at NASA, at a time when we really needed it most."

Shadow of the Shuttle Loss

Many NASA employees and many citizens were still reeling from the loss of the space shuttle Columbia and her crew of seven astronauts in February 2002, and the rovers were medicine for melancholy to be sure. "[MER] was an accomplishment that positively reassured us that we are and continue to do great things and are an extraordinary agency for all those purposes," said O'Keefe, who actually inherited both the 'culture' that led to the loss of the Columbia and her crew, as well as the team who put the rovers on Mars when he took over the administrative reins of the space agency three years ago.

Recalling that he and the team had toasted with champagne at the press conferences following the landings, something that apparently makes government lawyers very nervous, O'Keefe let on that the two empty bottles are currently on display in his office. "I'm as proud of [those] as a lot of the other mementos around the office, and as a matter of fact maybe more so because it was one of the most fulfilling days I have ever had in public service. It was an extraordinary experience. *Spirit*'s successful landing as well as the fantastic work of the orbiting MGS [Mars Global Surveyor] help set the table for President [George W.] Bush's confident announcement 11 days later on a bold vision for human and robotic exploration throughout the solar system. [MER] was a perfect precursor opportunity for exactly that, and we never planned it that way."

For the MER team—working on alternating shifts and living on Mars time—the stress experienced at the landings have continued through operations. "*Spirit* started out shortly after landing with the file management system anomaly," Erickson recalled. "And then we've had the challenge of going over two and a half kilometers up to the Columbia Hills. We went up to Bonneville Crater and then made the big drive to the West Spur of the Columbia Hills. *Opportunity* started us off with its own problem, a shoulder joint heater on the instrument arm was stuck on. We quickly learned how to work around it, but then it gave us the opportunity to practice driving on the steep, steep slopes inside Endurance Crater. That was a challenge in and of itself that we weren't really prepared for."

Spirit and *Opportunity* trekked onward, troopers through it all, and they are roving now, working right through the accolades, the memories, and the celebrations being held down on Earth. *Spirit* is carrying on with her explorations of the Columbia Hills within the Gusev Crater, climbing up Cumberland Ridge to a crest called Larry's Lookout on the way to the top of Husband Hill. "In December, we discovered a completely new type of rock in Columbia Hills, unlike anything seen before on Mars," informed Squyres.

At Meridiani Planum, *Opportunity* is checking out the heat shield which she jettisoned on her entry and wound up breaking into two pieces as it hit the ground. The study could reveal something to engineers about robustness of their designs. Rover team members hope to determine, for example, how deeply the atmospheric friction charred the protective layer. "Both pieces of the wreckage have great opportunities for close-up operations and should be able to tell us a great deal about how well it survived the impact with the atmosphere and maybe the impact with the surface," explained Erickson. "With luck, our observations could give us great opportunities to improve the chances for future missions when they enter the Martian atmosphere or any other planetary atmosphere."

Pushing the Limits

The plan for now is to drive the rovers "as hard as possible until they break down," Squyres offered. Despite the $3 million a month cost, the rovers probably will have the chance to shop for rocks 'til they drop. "It seems inconceivable to me that if the rovers are still producing good science that we will not find a way to keep them going," said Firouz Naderi, manager of NASA's Mars Exploration Program, at JPL. "We will manage somehow."

How long could they go, theoretically? No one really knows and no one right now is betting much on when exactly either rover will break down, because the twin robot field geologists continue to astound even their designers with how well they continue operating. "The rovers are in great shape for their age right now," reported Erickson. "They are continuing to set records. *Spirit* has driven more than 4 kilometers (2.5 miles) and *Opportunity* has completed 2 kilometers. Together they have returned 62,000 images and 86 gigabits of additional science data."

Spirit and *Opportunity* have also roved through the worst of the Martian winter with flying colors, and spring is on the horizon. Both rovers are in "strong positions" to continue exploring, Erickson said. "I can't tell you a timetable and in fact bad things could happen to us at any time," he cautioned. "Random part failures could lose emissions tomorrow. In addition to that, mild dust storms are kicking up now around the equatorial regions where the rovers are and that may impede the rovers' ability to collect solar power even though their arrays have remained cleaner than anticipated. "But as long as we have them we're going to keep using them to the best of their ability," he assured.

Beyond being the darlings of the day, *Spirit* and *Opportunity* are the most heralded robots of 2004. In December, the journal *Science* honored the MER mission for producing the "Breakthrough of the Year," for the rovers' "profound implications for society and the advancement of science." ["On Mars, A Second Chance for Life," *Science* 2004 306: 2010–2012].

"It's very difficult even from the distance of a year to put the findings of this mission into their proper perspective," Squyres said, as he was officially handed the floor. "To me, the most remarkable story of today, is that the rovers are still going and they are still making remarkable scientific discoveries," he added, anxious to share some of the MER mission's most recent findings.

Discovering New Materials

"*Spirit* has moved into something totally new," Squyres continued without missing a beat. "We have suddenly, in just the last few weeks, come into a completely different geologic material unlike anything seen on Mars before. This came as a complete surprise to us. I lost a bet on this one," he added, drawing laughter.

The MER team found its latest surprise in a rock called Wishstone, and another dubbed Wishing Well that *Spirit* is examining now. "We did some measurements down in the interior of [Wishstone] and what we found was a rock that was really different from anything we've ever seen on Mars before, even in the Columbia Hills, we have not seen rocks like this before," Squyres explained.

Wishstone is a rock composed of grains—"one a geologist would call an iclastic rock," Squyres elaborated. "But the grains have enormous ranges of sizes—it's a poorly sorted rock. Now, there are many geologic processes that make rocks out of grains that will make all the grains the same sizes . . . but this isn't like that. This [rock] has grains of all sizes from very small to very large and that tends to indicate that it was put into place in a very high-energy environment. It happened fast. We've seen this before in the rocks of the Columbia Hills, but this is another, a different indication of that. Some of these grains are very angular, look sort of busted up, which may also speak to a rapid violent process involved in forming this rock."

The two obvious candidates, Squyres continued, are some kind of geologic explosion or an impact process. "It's difficult to tell the two apart, but probably one of those two was responsible for making this rock." The "weird stuff" emerges in the chemistry.

The first message is that [this rock] is dramatically different from anything we've seen before. The second things is . . . this rock is chock full of phosphorous . . . a much higher phosphorus content than anything we've seen on Mars before."

There are a couple possible explanations for the rich phosphorous content, Squyres said. "One possibility is that the igneous rock itself was rich in phosphorus to start with—it was a primary mineral in the rock. The other possibility is that it was a phosphate that was deposited from water. That's a possibility we are pursuing. If it is what happened, then it speaks of a water chemistry dramatically different from what we saw just 500 hundred meters away on the West Spur . . . [where] we saw chlorine, sulfur, bromine—elements that are way down in this rock.

"Somehow [Wishstone] is telling us something about a dramatic difference in the water chemistry, somehow the water chemistry changing with position or with time," Squyres explained. "There are some ways of making phosphates that involve water; others do not. As we look at a family of these rocks together, we should be able to test between those possibilities.". . .

Studying the Heat Shield

As *Opportunity* has roved south from Endurance Crater, up to and around the two pieces of her heat shield during the last week, the scientists have been having fun. "We get to sit back and let all the engineers do the really hard thinking and we just take pictures and make the engineers happy," explained Squyres. "It's a complete reversal of role—and it's kind of cool."

Opportunity's current objective is to get up and image the debris with the panorama camera (PanCam), as well as the microscopic imager (MI) to determine how it handled the blast through the atmosphere. The rover is driving around the back side of the flank portion of debris to acquire images of the heat shield's perimeter. Once her work there is done, she will take a long drive to the south, heading for an area called Etched Terrain, which the science team homed in on for study months ago.

"It's going to be a connect the dots affair," said Squyres, as he indicated the number of small craters on an image of *Opportunity*'s landing area taken by the Mars orbital camera (MOC) aboard the orbiting *Mars Global Surveyor*, and projected on the movie screen behind him. "We're going to hop-scotch from crater to crater—and our first key objective is a circular feature we call Vostok. It's sort of crater-like in appearance but its funny looking . . . we think it's a strangely eroded impact crater.

No matter what happens from this point forward, *Spirit* and *Opportunity* can retire as champions at any point from here on out. They have given scientists enough data to keep them busy for decades. Their riding-off-into-the-Martian-sunset ending is already written. "They could die tomorrow, a month from now, or a year from now," summed up Manning. "Whatever happens, it has been a fantastic journey for these two rovers."

"For all of us, this mission with *Spirit* and *Opportunity* has been—in the very literal sense of the phrase—the adventure of a lifetime and I know I speak for the whole team when I say how fortunate we feel to have been able to be a part of it," a humbled Squyres said. "I think by any reckoning, the legacy, the lasting legacy of this mission is going to turn out to be the recognition that our sister planet, Mars, once had habitable conditions on its surface. What that meant for the origin of life and what it meant for the evolution of life, that's for future Mars programs to determine . . . but I think we have set the future Mars program a direction and a goal to pursue."

A Collaborative Success

While NASA has had plenty of opportunity to review and analyze causes of failures in the recent past, Elachi, Squyres, Erickson, Manning and others on the MER team had no trouble identifying the root cause of the MER mission's success. In two words: teamwork and resources.

"The biggest thing that we have is the teamwork and the people involved at JPL—both the people who did the design the operations, the original thoughts, the team that came with Dr. Squyres from across the country and across the world all working together—that was what led to the success of this mission," offered Erickson.

"This was just an extraordinary group of individuals that did these rovers," Squyres concurred. "When we needed resources we got them—from the agency, from the Jet Propulsion Laboratory, and we got the people we needed and these guys did things right. It's as simple as that."

The Search for Exoplanets

By Jeff Foust

For centuries people have wondered if planets similar to Earth exist elsewhere. Since 1991 astronomers have confirmed the existence of dozens of massive planets orbiting other stars. Most are gas giants with many times the mass of Jupiter, the largest planet in the solar system. Since the turn of the century, somewhat smaller planets, similar in size to Neptune, have also been discovered, but so far no Earth-like exoplanet (as planets orbiting other stars are termed) has been detected. Terrestrial exoplanets may exist, however.

In the following selection Jeff Foust discusses the search for exoplanets. Noting that from the edge of the solar system even the largest planets look tiny, Foust makes the point that exoplanets cannot be directly observed by Earth-based telescopes. That remains largely true. However, in early 2005 a European consortium using a large telescope in Chile produced what it says is the first image of an exoplanet, a planet that is more than ten times the mass of Jupiter. The main method by which astronomers detect exoplanets is by carefully measuring the wobble that the planets cause in the stars they orbit. There are multiple ways to measure this gravitationally induced wobble, and Faust reports that new and better techniques are being developed. Another method of searching for exoplanets involves measuring the dimming of a star as a planet passes in front of it. The National Aeronautics and Space Administration plans to launch a space-based system to look for Earth-like planets around 2011. Jeff Foust is the editor and publisher of the *Space Review*. He also operates the Spacetoday.net Web site.

Jeff Foust, "Finding Those Pale Blue Dots," *Ad Astra,* vol. 14, January/February 2002, pp. 19–23.

■n 1990, the Voyager 1 spacecraft swung its camera towards the Sun and took a set of images, creating the first "family portrait" of our solar system. From a distance of 4 billion miles even the largest planets, Jupiter and Saturn, appeared as little more than faint points of light. The Earth was nearly lost in the glare of the Sun, visible only as a "pale blue dot," as described by the late [astronomer] Carl Sagan.

Today astronomers turn their telescopes towards other stars in a search for planets orbiting those stars. Their ultimate quest is not just to discover other planets, but to find other "pale blue dots": other Earths, worlds that could be home to life. Such a search is far more difficult than trying to view the Earth from the edge of the solar system, and pushes our technology to the limits. However, the ultimate outcome of such a search—the discovery of an Earthlike planet around another star—would be one of the greatest discoveries in the history of astronomy.

First Finds of Exoplanets

The first confirmed discovery of an extrasolar planet, or "exoplanet," was made in 1991. Alexander Wolszczan, a Penn State University astronomer, found three small planets orbiting the pulsar PSR1257+12. By measuring tiny variations in the regular pattern of signals from the pulsar, Wolszczan determined that two of the planets weighed about the same as the Earth, and a third about the same as the Moon. While interesting, most astronomers consider the planetary system PSR1257+12 an anomaly: the planets may have formed after the supernova explosion that created the pulsar. Moreover, any planet orbiting a pulsar would unlikely be Earthlike in any way, and in fact, would be among the most inhospitable places for life.

In 1995 the first planet around a sunlike star was discovered. Michel Mayor and Didier Queloz of the Geneva Observatory detected a planet around 51 Pegasi, a star 50 light years from Earth. At first some astronomers thought Mayor and Queloz had made a mistake: the planet, with an estimated mass about half that of Jupiter, was orbiting just 7.5 million kilometers away, completing one orbit around the star in just over four days. However, the existence of the planet was soon confirmed by American astronomers Geoffrey Marcy and Paul Butler. This discovery opened the floodgates for many more, most by either Marcy's group or Mayor's. By

November 2001 astronomers had discovered 76 exoplanets.

Because the technology does not yet exist to directly observe exoplanets through telescopes, astronomers rely on a number of techniques to detect exoplanets indirectly. The most commonly used method is the "radial velocity" technique. The gravity of an orbiting exoplanet tugs on its parent star, creating a very slight but regular wobble. The wobble cannot be seen by Earth-based telescopes, but the effect of the wobble on the star's spectra can be observed. As the star wobbles towards and away from the Earth, the wavelength of light from the star changes according to the Doppler effect, in much the same way the pitch of a horn is higher from an approaching car and lower from one moving away. By studying a star with a spectrograph over long periods of time, astronomers can measure the magnitude of the star's wobble, called its radial velocity, and the period of the wobble, enough to deduce the location and mass of the exoplanet that causes it.

The radial velocity technique has been very successful: Marcy and Mayor, among others, have used it to discover virtually all the exoplanets known to date. One disadvantage of the method, though, is that it is biased towards planets that can generate the largest wobbles in stars: these tend to be large planets that either closely orbit their stars or planets which occupy more distant, but eccentric, orbits that swing close to the star for a part of each orbit. Neither type of planet is particularly hospitable for life, but in the first few years after the discovery of the planet around 51 Pegasi, most new exoplanets tended to be one type or the other.

Whole Solar Systems Discovered

Those statistics are starting to change, though, as astronomers gather additional data covering longer periods of time. In 1999 astronomers discovered the first extrasolar solar system: three planets around the star Upsilon Andromedae. One planet is very close to the star but the other two planets are more distant. By 2001 seven such solar systems had been discovered, each with at least two gas giant planets. In August 2001 astronomers found that the star 47 Ursae Majoris has two planets orbiting the star in circular orbits where the asteroid belt would be in our solar system. Two months later astronomers found two more extrasolar planets around other stars, each in near circular orbits near where Mars would be in our solar system.

Astronomers credit the discovery of more "normal" planets to the large amounts of data they have discovered, and improvements in the quality of the data. "As our sensitivity improves we are finally seeing planets with longer orbital periods, planetary systems that look more like our solar system," explained Debra Fischer, an astronomer at the University of California Berkeley and a colleague of Marcy and Butler.

"Most of the planetary systems we've found have looked like very distant relatives of the solar system, no family likeness at all," says Steve Vogt of the University of California Santa Cruz. "Now we're starting to see something like second cousins. In a few years' time we could be finding brothers and sisters."

New Methods Needed

While astronomers come closer to finding siblings of gas giants like Jupiter and Saturn around other stars, even distant relatives of Earth remain elusive. The radial velocity technique cannot detect the microscopic wobbles that an Earth-sized planet creates in a star. Today, astronomers can detect wobbles as small as 3 meters per second using the radial velocity technique. This is more than enough to detect Jupiter-sized planets—Jupiter creates a 12 m/s wobble in the Sun—but falls well short of being able to find terrestrial worlds.

The radial velocity technique can be improved to some degree. Marcy's group is trying to raise $5 million to build a two-meter telescope dedicated to extrasolar planet searches. "With a dedicated telescope we could begin to detect much lower mass planets—perhaps as low as 20 Earth masses," says Fischer. However, even this limit is far more massive than the Earth. Finding terrestrial worlds will require new methods and new telescopes.

One alternative to the radial velocity technique is to directly observe the wobble of the stars. To do this requires very accurate astrometry, or measurements of the positions of stars. The technology for doing this is being developed today, using a procedure called interferometry: combining light from distant telescopes in such a way to simulate a single, very large telescope. The Palomar Testbed Interferometer (PTI), located at Palomar Observatory in California, is an effort by JPL to test the astrometric technologies needed to perform extrasolar planet searches. The technology from PTI is being applied to a larger interferometer at the Keck Observatory in Hawaii. The Keck Interferometer should be able to observe wobbles created by planets as small as Uranus, about 15 times the mass of the Earth. A similar interferometer system is under development at the European Southern Observatory in Chile.

Astrometry can also be performed from space, either using single telescopes or interferometer systems. The Full-sky Astrometric Mapping Explorer (FAME), a NASA [National Aeronautics and Space Administration] mission scheduled for launch in 2004, will use a single telescope to measure the positions of 40 million stars. FAME will be able to detect very large planets, at least twice the mass of Jupiter, and will focus on looking for even larger "super Jupiters" and brown dwarfs. StarLight, formerly known as Space Technology 3, is a NASA mission scheduled for launch in 2005 that will be the first test of an interferometer in space. StarLight will not perform any new science, but will test the technologies needed for future space interferometers.

The ultimate in astrometric planet searches will come with the launch of the Space Interferometry Mission (SIM) in 2009. SIM will use three pairs of small telescopes mounted on a boom 10 meters long to measure the wobbles of stars with unprecedented resolution. SIM will be able to detect wobbles created by planets only five to ten times the mass of the Earth orbiting nearby Sunlike stars. NASA is developing SIM with Lockheed Martin and TRW.

Planets in Transit

Measuring, directly or indirectly, the wobbles of stars is not the only way to detect exoplanets. Another technique is to look for transits, when an exoplanet passes directly between the Earth and the star it orbits. During the transit the planet blocks a tiny fraction of light from the star—around 1% for a gas giant like Jupiter, and 0.01% for an Earthlike planet—which can be detected by telescopes on Earth. Measuring this drop, and the length of the transit, allows astronomers to measure the size of the planet and its orbit. This method has already been used to confirm the existence of a planet that was first discovered using the radial velocity technique. A key advantage of transits is that they can be used to discover planets far smaller than those detectable using radial velocity or astrometry, including planets the size of Earth.

Transits can also be used to study the atmospheres of planets. In late November 2001 astronomers reported the first discovery of an atmosphere around an exoplanet, using HD 209458b, the one exoplanet to date known to transit across the face of its star. Astronomers were able to detect sodium absorption lines in spectra of the star as the planet transited; these lines, they concluded, came from the small fraction of light that passed through the planet's atmosphere. While sodium is a "tracer gas" that makes up only a few parts per million of the atmosphere, it does provide evidence that the atmosphere exists, and paves the way for future studies that may be able to look for other components of exoplanet atmospheres.

A problem with transit photometry, though, is that transits are not common: the planet must be perfectly aligned between the Earth and star for a transit to be visible. To use this method to discover planets, rather than to confirm planets found by other means, requires dedicated telescopes that can observe large numbers of stars simultaneously over long periods of time. Several spacecraft missions are being developed to accomplish this. Kepler, a mission recently selected by NASA's Discovery Program, will fly a wide-angle telescope capable of observing 100,000 stars simultaneously, looking for brief, periodic drops in brightness that could be caused by a transiting planet. Kepler is designed to look for planets the size of Earth, although it is capable of detecting a planet as small as Mercury. Kepler is scheduled for launch in 2005.

Other missions will also use transit photometry to look for planets. The Microvariability and Oscillation of Stars (MOST) mission, the first Canadian space telescope, will use a 15-centimeter telescope to, among other things, look for brief variations in starlight that could be caused by a transiting planet. MOST is scheduled for launch in October 2002 on a Russian rocket booster. The French Corot mission, scheduled for launch in 2004, will also look for exoplanet transits, but only as a secondary component of its primary mission to study the interiors of stars. Eddington, a European Space Agency (ESA) mission scheduled for launch in 2008, will spend three years looking at a single star field in an effort to detect planetary transits.

Searching for Earthlike Planets

Advances in radial velocity, astrometry, and transit techniques will make it possible in the coming decade to discover planets possibly as small as Earth. However, in most cases these techniques will tell us nothing about the planets themselves: there is no way for them to tell the difference between a terrestrial planet hospitable to life—another Earth—and a terrestrial planet utterly inhospitable to life—another Venus. To determine which exoplanets are truly Earthlike will require a new generation of space observatories able to directly observe these worlds.

NASA is proposing one such mission, called Terrestrial Planet Finder. TPF will build on the space interferometry techniques tested by StarLight and SIM by flying several telescopes on separate spacecraft in formation and combining their light to create a single image. Current plans call for TPF to use four telescopes, each 3.5 meters in diameter (more than 50% larger than the Hubble Space Telescope), spread out by as much as one kilometer. In this configuration TPF would be able to directly observe an exoplanet the size of the Earth around a nearby star in as little as two hours of observations, and obtain a detailed spectrum of the planet in two days. NASA is planning on a 2011 launch of TPF, but given that this is just two years after the launch of SIM, it's quite likely TPF will not be launched until later in the 2010s.

ESA is also considering its own version of TPF, named Darwin. This mission will use six telescopes, each 1.5 meters across, combined using interferometry much like TPF. Darwin will be able to directly detect terrestrial planets and take spectra of them. ESA is planning a 2014 launch of Darwin, possibly in cooperation with Russia or Japan. The agency is also considering joining forces with NASA and developing a single TPF/Darwin mission.

Picking Out Hospitable Targets

Once terrestrial planets can be directly observed, there are a number of ways to determine which may be like the Earth. The key method is through spectroscopy, looking at specific wavelengths of light for absorption lines by key elements and compounds. At infrared wavelengths scientists will be able to see evidence of water, methane, ozone, and carbon dioxide: all critical to life on Earth. An exoplanet whose spectra contains those signatures, and at the right temperature, is likely to be an Earthlike planet.

Princeton University scientists outlined another way to identify Earthlike planets in a paper published in the journal *Nature* in August [2001]. They found that the albedo, or reflectivity, of the Earth changes considerably over the course of a single day as the amount of clouds, water, and land that is visible changes. Their model of Earth showed that its brightness changed by as much as 20% over a single day, compared to virtually no variations in Venus and Mars. Thus, by simply measuring the brightness of an exoplanet over the course of a few days, scientists may be able to tell the difference between an Earthlike planet and a less hospitable world.

Eric Ford, the lead author of the *Nature* paper, thinks that these two techniques can complement and assist each other. "Spectroscopy will be necessary to uniquely identify some molecules in the atmosphere," he says. "Photometric variability can tell you about the rotation, geography, and climate of a planet."

Of course, astronomers will not stop when they discover an Earthlike world: NASA already has on the drawing boards proposals for giant space observatories that would be able to take images of these worlds at high enough resolutions to discern continents and oceans. However, simply the knowledge that there is another planet out there like the Earth—another pale blue dot—may be one of the most profound discoveries ever made.

CHAPTER 4

Modern Controversies in Astronomy

Extraterrestrial Intelligence Is Likely to Exist

By Annette Foglino

The idea that there might be intelligent life on other worlds caught on in the nineteenth century, when astronomer Percival Lowell claimed (erroneously) that he had detected a civilization on Mars. Since the 1960s, the effort to detect intelligent life on other planets has become known as SETI—the Search for Extraterrestrial Intelligence. The following selection presents interviews with scientists conducted by Annette Foglino. During the interviewing process, she spoke with scientists holding a range of views on the likelihood of extraterrestrial life and the chances that any such life might be intelligent. The interviews excerpted in this viewpoint are with those scientists who believe intelligent life exists. Astronomer Frank Drake predicts that by 2089 other civilizations will have been discovered. Astronomer Jill Tarter is also confident that intelligent extraterrestrial life exists. She thinks that such a life-form is likely to have a body and head somewhat similar to humans. Radio astronomer Robert Dixon argues that there is nothing especially unique about the earth and sun, leading him to conclude that the conditions that make life on Earth possible also exist elsewhere in space. Physicist Paul Horowitz expresses certainty that intelligent life exists beyond Earth. Searching for it is worthwhile, he argues, because of the potential to learn new things. Annette Foglino, an award-winning journalist, is a former writer and reporter for *Life* who has also written for numerous other publications.

From the malevolent Martians of H.G. Wells's *The War of the Worlds* to the hirsute hunks in this summer's [1989] hit film *Earth Girls Are Easy*, extraterrestrials have long been the subject of earthly speculation. But until now, serious scientific efforts to make contact with other intelligent beings have been limited to a few dozen sporadic and poorly funded projects. In 1960 astronomer Frank Drake launched the first, dubbed OZMA, using a special hydrogen radio frequency to listen in on "transmissions" within our galaxy. Drake failed to detect any, although he continues to try. In 1985 Harvard University began using META (Megachannel Extraterrestrial Assay) to monitor more than eight million space radio channels.

Next spring [1990] the search will move into high gear when the first space telescope is put into orbit by the shuttle astronauts, allowing astronomers to peer into the heavens with more clarity than ever before. Meanwhile, Congress is considering a 10-year, $100 million NASA [National Aeronautics and Space Administration] program. Ten billion times more powerful than META, the SETI (Search for Extraterrestrial Intelligence) project will use 15 million channels and target specific stars as well as eavesdrop on the whole sky. The system is to be launched on Columbus Day 1992, 500 years after the explorer touched land in the New World.

Life [magazine] has polled 35 leading scientists in the field on whether life exists in space. Twenty-six said they believed it does in some form in our galaxy, four said there is little chance, and five reserved judgment pending further evidence. On the following pages, some of these experts give their reasons and reflect on what we might encounter if our great reach outward is met one day by an alien grasp. . . .

Frank Drake, president of the SETI Institute, is an astrophysicist at the University of California at Santa Cruz:

The chances are very high—I'd say 100 percent. We know a great deal about the processes that took place in the solar system and of life on earth, and they were all completely normal processes. No freak events were required. So one would expect that the same sequence of events has occurred at least in a few places—and probably in many, many places.

By examining the moons of planets like Jupiter and Saturn, we might be able to trace the chemistry of creation and the formation of the molecules of life in situations like those on primitive earth. We've seen that there are many chemical pathways that lead to the creation of living things. And we have seen in meteorites and interstellar clouds the very molecules that can react to produce the molecules of life.

This is going to be the hot topic of the next century. By 2089 we will have succeeded in learning about other civilizations. Do they typically colonize space, their own planetary systems? How do they cope with increases in population? . . .

Jill Tarter is project scientist for NASA's Microwave Observing Project at California's Ames Research Center:

The odds are overwhelmingly in favor of intelligent life existing somewhere else in our galaxy. From studying fossil records, there doesn't seem to be anything extraordinarily original about our planet. First of all, life got started on its surface almost as soon as it could—there is a very short time span between the formation of a cool terrestrial surface, one that is rich with water, and the appearance of the oldest microfossil evidence of life. This tells me that the process that led to the organization of the first self-replicating cell wasn't terribly complicated or improbable. Given similar conditions, one could expect similar results. If the path toward life was terribly complicated and improbable, then it seems to me that it would have taken a substantial amount of time to have happened on this planet. But it didn't. To me that implies, in some sense, that life is easy.

The road from microbes to man was long and convoluted and it might easily have been sidetracked. But I think that it will eventually turn out that our theorists are right: Planets around other stars are the rule rather than the exception. And with 400 billion stars in our galaxy and 100 billion galaxies in the universe, the odds are good enough that similar outcomes will have transpired elsewhere. There is a lot to be said for form following function. Things with 13 legs and one-sided fish don't seem to have survived. I would expect that any other intelligent being would be bilaterally symmetrical. It would have to have some sort of head or a special place in its body that protects its central nervous system. If it were to develop a technology, it would need manipulative appendages: arms, legs, tail, something with which to modify its environment. A universe populated by microbes would be fairly lonely. Maybe this is a fundamental question haunting the human spirit, since we don't like vast spaces where we stand alone. . . .

Robert Dixon, radio astronomer, is director of Ohio State's SETI program:

I think it's very likely. Everywhere astronomers look in the universe, we see the same physical laws, the same chemical elements and properties. There is nothing overtly odd about our sun or earth as far as we can discern. There are likely planets everywhere, but we can't see them because our telescopes aren't big enough.

The important point is that we're not trying for two-way communication at this stage. When we watch television, we're engaging in one-way communication. We can't answer back. When we read a book by an ancient Greek philosopher, not only is that person dead but his entire civilization is dead as well. Yet this doesn't deter us. We still find his information useful. What we're doing now is searching for life elsewhere to answer the basic question: "Are we alone?" And if we're lucky, we'll discover the equivalent of the *Encyclopedia Galactica.* It's an information-gathering process rather than a communication process.

There are some people who say that we should never answer— even if we do get a signal—because then we'll give away our existence and the evil galactic monsters will swoop down and enslave us. The problem with this approach is the cat is already out of the bag. We've been sending our TV shows out into space at the speed of light for years now.

Paul Horowitz, a physicist at Harvard University, helped set up META, the most sophisticated SETI listening post in operation: Life out of earth exists, absolutely guaranteed. I'll give you any odds. As far as intelligent life in the universe, I am essentially certain. And as for intelligent life in our galaxy, I would have to say that it is likely.

My opinion is based on a lot of facts. Our sun is as ordinary a star as it is possible to be. Life arose on earth through natural processes almost as swiftly as it could. As soon as the oceans formed, chemical evolution gave rise to life's building blocks. We also see many examples of organic molecules in space, in meteorites. We see evidence of the beginnings of planetary systems. As soon as we've achieved enough sensitivity with our instruments of detection, we will discover life.

When people ask, "Why should we bother communicating with these folks out there?" I answer: "We've only reached this stage by being curious." Extraterrestrials that we might encounter would be more advanced than we are—so if we keep up the search, we'll eventually learn things we don't know.

Extraterrestrial Intelligence Is Probably Exceedingly Rare

By Peter D. Ward and Donald Brownlee

In the following selection geologist Peter D. Ward and astronomer Donald Brownlee dispute the idea that intelligent life is widespread in the universe. On the contrary, they argue, while life may be commonplace, intelligent life is probably exceedingly rare. Indeed, taking into account the many favorable circumstances necessary for complex life to survive, they wonder whether Earth might not be the *only* outpost of animal life in the universe. Among the factors that make intelligent life on this planet possible are Jupiter, which blocks all but a few large meteors and comets from striking Earth, and the sun, which provides a steady stream of energy. Without these and many other favorable circumstances, they argue, life on Earth would never have gotten past the bacterial stage. The authors contend that the confluence of factors necessary to the development of intelligent life would be unlikely to arise elsewhere in the universe. Peter D. Ward is a professor of geological sciences and zoology as well as curator of paleontology at the University of Washington in Seattle. Don Brownlee is a professor of astronomy at the same university.

On any given night, a vast array of extraterrestrial organisms frequent the television sets and movie screens of the world. From *Star Wars* and *Star Trek* to *The X-Files*, the message is clear: The Universe is replete with alien life forms that vary

Peter D. Ward and Donald Brownlee, *Rare Earth: Why Complex Life Is Uncommon in the Universe.* New York: Copernicus, 2000. Copyright © 2000 by Peter D. Ward and Donald Brownlee.

widely in body plan, intelligence, and degree of benevolence. Our society is clearly enamored of the expectation not only that there is *life* on other planets, but that incidences of *intelligent* life, including other civilizations, occur in large numbers in the Universe.

This bias toward the existence elsewhere of intelligent life stems partly from wishing (or perhaps fearing) it to be so and partly from a now-famous publication by astronomers Frank Drake and Carl Sagan, who devised an estimate (called the Drake Equation) of the number of advanced civilizations that might be present in our galaxy. This formula was based on educated guesses about the number of planets in the galaxy, the percentage of those that might harbor life, and the percentage of planets on which life not only could exist but could have advanced to exhibit culture. Using the best available estimates at the time, Drake and Sagan arrived at a startling conclusion: Intelligent life should be common and widespread throughout the galaxy. In fact, Carl Sagan estimated in 1974 that a million civilizations may exist in our Milky Way galaxy alone. Given that our galaxy is but one of hundreds of billions of galaxies in the Universe, the number of intelligent alien species would then be enormous.

Concealed Assumptions

The idea of a million civilizations of intelligent creatures in our galaxy is a breathtaking concept. But is it credible? The solution to the Drake Equation includes hidden assumptions that need to be examined. Most important, it assumes that once life originates on a planet, it evolves toward ever higher complexity, culminating on many planets in the development of culture. That is certainly what happened on our Earth. Life originated here about 4 billion years ago and then evolved from single-celled organisms to multicellular creatures with tissues and organs, climaxing in animals and higher plants. Is this particular history of life—one of increasing complexity to an animal grade of evolution—an inevitable result of evolution, or even a common one? Might it, in fact, be a very rare result? . . .

We will argue that not only intelligent life, but even the simplest of animal life, is exceedingly rare in our galaxy and in the Universe. We are not saying that *life* is rare—only that *animal* life is. We believe that life in the form of microbes or their equivalents is very common in the universe, perhaps more common than even

Drake and Sagan envisioned. However, *complex* life—animals and higher plants—is likely to be far more rare than is commonly assumed. We combine these two predictions of the commonness of simple life and the rarity of complex life into what we will call the Rare Earth Hypothesis. In the pages ahead we explain the reasoning behind this hypothesis, show how it may be tested, and suggest what, if it is accurate, it may mean to our culture.

The search in earnest for extraterrestrial life is only beginning, but we have already entered a remarkable period of discovery, a time of excitement and dawning knowledge perhaps not seen since Europeans reached the New World in their wooden sailing ships. We too are reaching new worlds and are acquiring data at an astonishing pace. Old ideas are crumbling. New views rise and fall with each new satellite image or deep-space result. Each novel biological or paleontological discovery supports or undermines some of the myriad hypotheses concerning life in the Universe. It is an extraordinary time, and a whole new science is emerging: astrobiology, whose central focus is the condition of life in the Universe. The practitioners of this new field are young and old, and they come from diverse scientific backgrounds. Feverish urgency is readily apparent on their faces at press conferences, such as those held after the Mars Pathfinder experiments, the discovery of a Martian meteorite on the icefields of Antarctica, and the collection of new images from Jupiter's moons. In usually decorous scientific meetings, emotions boil over, reputations are made or tarnished, and hopes ride a roller coaster, for scientific paradigms are being advanced and discarded with dizzying speed. We are witnesses to a scientific revolution, and as in any revolution there will be winners and losers—both among ideas and among partisans. It is very much like the early 1950s, when DNA was discovered, or the 1960s, when the concept of plate tectonics and continental drift was defined. Both of these events prompted revolutions in science, not only leading to the complete reorganization of their immediate fields and to adjustments in many related fields, but also spilling beyond the boundaries of science to make us look at ourselves and our world in new ways. That will come to pass as well in this newest scientific revolution, the Astrobiology Revolution of the 1990s and beyond. What makes this revolution so startling is that it is happening not within a given discipline of science, such as biology in the 1950s or geology in the 1960s, but as a convergence of widely different scientific disciplines: astronomy, biology, paleontology, oceanography, microbiology, geology, and genetics, among others.

Thinking Big in Biology

In one sense, astrobiology is the field of biology ratcheted up to encompass not just life on Earth but also life beyond Earth. It forces us to reconsider the life of our planet as but a single example of how life might work, rather than as the only example. Astrobiology requires us to break the shackles of conventional biology; it insists that we consider entire planets as ecological systems. It requires an understanding of fossil history. It makes us think in terms of long sweeps of time rather than simply the here and now. Most fundamentally, it demands an expansion of our scientific vision—in time and space.

Because it involves such disparate scientific fields, the Astrobiology Revolution is dissolving many boundaries between disciplines of science. A paleontologist's discovery of a new life form from billion-year-old rocks in Africa is of major consequence to a planetary geologist studying Mars. A submarine probing the bottom of the sea finds chemicals that affect the calculations of a planetary astronomer. A microbiologist sequencing a string of genes influences the work of an oceanographer studying the frozen oceans of Europa (one of Jupiter's moons) in the lab of a planetary geologist. The most unlikely alliances are forming, breaking down the once-formidable academic barriers that have locked science into rigid domains. New findings from diverse fields are being brought to bear on the central questions of astrobiology: How common is life in the universe? Where can it survive? Will it leave a fossil record? How complex is it? There are bouts of optimism and pessimism; E-mails fly; conferences are hastily assembled; research programs are rapidly redirected as discoveries mount. The excitement is visceral, powerful, dizzying, relentless. The practitioners are captivated by a growing belief: Life is present beyond Earth.

Early Disappointments

The great surprise of the Astrobiology Revolution is that it has arisen in part from the ashes of disappointment and scientific despair. As far back as the 1950s, with the classic Miller-Urey ex-

periments showing that organic matter could be readily synthesized in a test tube (thus mimicking early Earth environments),
scientists thought they were on the verge of discovering how life
originated. Soon thereafter, amino acids were discovered in a
newly fallen meteorite, showing that the ingredients of life occurred in space. Radio-telescope observations soon confirmed this,
revealing the presence of organic material in interstellar clouds. It
seemed that the building blocks of life permeated the cosmos.
Surely life beyond Earth was a real possibility.

When the Viking I spacecraft approached Mars in 1976, there
was great hope that the first extraterrestrial life—or at least signs
of it—would be found. But Viking did *not* find life. In fact, it
found conditions hostile to organic matter: extreme cold, toxic soil
and lack of water. In many people's minds, these findings dashed
all hopes that extraterrestrial life would ever be found in the solar
system. This was a crushing blow to the nascent field of astrobiology. . . .

Unless it can be shown that life can form, as well as live, in extreme environments, there is little hope that even simple life is
widespread in the Universe. Yet here, too, revolutionary new findings lead to optimism. Recent discoveries by geneticists have
shown that the most primitive forms of life on Earth—those that
we might expect to be close to the first life to have formed on our
planet—are exactly those tolerant life forms that are found in extreme environments. This suggests to some biologists that life on
Earth *originated* under conditions of great heat, pressure, and lack
of oxygen—just the sorts of conditions found elsewhere in space.
These findings give us hope that life may indeed be widely dis-
tributed, even in the harshness of other planetary systems.

Fossil Record

The fossil record of life on our own planet is also a major source
of relevant information. One of the most telling insights we have
gleaned from the fossil record is that life formed on Earth about
as soon as environmental conditions allowed its survival. Chemical traces in the most ancient rocks on Earth's surface give strong
evidence that life was present nearly 4 billion years ago. Life thus
arose here almost as soon as it theoretically could. Unless this occurred utterly by chance, the implication is that nascent life itself
forms—is synthesized from nonliving matter—rather easily. Per-

haps life may originate on *any* planet as soon as temperatures cool to the point where amino acids and proteins can form and adhere to one another through stable chemical bonds. Life at this level may not be rare at all.

The skies too have yielded astounding new clues to the origin and distribution of life in the Universe. In 1995 astronomers discovered the first extrasolar planets orbiting stars far from our own. Since then, a host of new planets have been discovered, and more come to light each year.

For a while, some even thought we had found the first record of extraterrestrial life. A small meteorite discovered in the frozen icefields of Antarctica appears to be one of many that originated on Mars, and at least one of these may be carrying the fossilized remains of bacteria-like organisms of extraterrestrial origin. The 1996 discovery was a bombshell. The President of the United States announced the story in the White House, and the event triggered an avalanche of new effort and resolve to find life beyond Earth. But evidence—at least from this particular meteorite—is highly controversial.

All of these discoveries suggest a similar conclusion: Earth may not be the only place in this galaxy—or even in this solar system—with life. Yet if other life is indeed present on planets or moons of our solar system, or on far-distant planets circling other stars in the Universe, what kind of life is it? What, for example, will be the frequency of *complex metazoans*, organisms with multiple cells and integrated organ systems, creatures that have some sort of behavior—organisms that we call animals? Here too a host of recent discoveries has given us a new view. Perhaps the most salient insights come, again, from Earth's fossil record.

New ways of more accurately dating evolutionary advances recognized in the Earth's fossil record, coupled with new discoveries of previously unknown fossil types, have demonstrated that the emergence of animal life on this planet took place later in time, and more suddenly, than we had suspected. These discoveries show that life, at least as seen on Earth, does not progress toward complexity in a linear fashion but does so in jumps, or as a series of thresholds. Bacteria did not give rise to animals in a steady progression. Instead, there were many fits and starts, experiments and failures. Although life may have formed nearly as soon as it could have, the formation of *animal* life was much more recent and protracted. These findings suggest that complex life is far more difficult to arrive at than evolving life itself and that it takes a much longer time period to achieve.

Threats to Survival

It has always been assumed that attaining the evolutionary grade we call animals would be the final and decisive step: that once this level of evolution was achieved, a long and continuous progression toward intelligence should occur. However, another insight of the Astrobiological Revolution has been that *attaining* the stage of animal life is one thing, but *maintaining* that level is quite something else. New evidence from the geological record has shown that once it has evolved, complex life is subject to an unending succession of planetary disasters that create what are known as mass extinction events. These rare but devastating events can reset the evolutionary timetable and destroy complex life, while sparing simpler life forms. Such discoveries again suggest that the conditions hospitable to the evolution and existence of *complex* life are far more specific than those that allow life's *formation*. On some planets, then, life might arise and animals eventually evolve—only to be quickly destroyed by a global catastrophe.

To test the Rare Earth Hypothesis—the paradox that life may be nearly everywhere but complex life almost nowhere—may ultimately require travel to the distant stars. We cannot yet journey much beyond our own planet, and the vast distances that separate us from even the nearest stars may prohibit us from ever exploring planetary systems beyond our own. Perhaps this view is pessimistic, and we will ultimately find a way to travel much faster (and thus farther), through worm holes or other unforeseen methods of interstellar travel, enabling us to explore the Milky Way and perhaps other galaxies as well.

Let's assume that we do master interstellar travel of some sort and begin the search for life on other worlds. What types of worlds will harbor not just life, but complex life equivalent to the animals of Earth? What sorts of planets or moons should we look for? Perhaps the best way to search is simply to look for planets that resemble Earth, which is so rich with life. Do we have to duplicate this planet exactly to find animal life, though? What is it about our solar system and planet that has allowed the rise of complex life and nourished it so well? Addressing this issue in the pages ahead should help us answer the other questions we have posed.

Steady Sun

If we cast off our bonds of subjectivity about Earth and the solar system, and try to view them from a truly "universal" perspective, we also begin to see aspects of Earth and its history in a new light. Earth has been orbiting a star with relatively constant energy output for billions of years. Although life may exist even on the harshest of planets and moons, animal life—such as that on Earth—not only needs much more benign conditions but also must have those conditions present and stable for great lengths of time. Animals as we know them require oxygen. Yet it took about 2 billion years for enough oxygen to be produced to allow all animals on Earth. Had our sun's energy output experienced too much variation during that long period of development (or even afterward), there would have been little chance of animal life evolving on this planet. On worlds that orbit stars with less consistent energy output, the rise of animal life would be far chancier. It is difficult to conceive of animal life arising on planets orbiting variable stars, or even on planets orbiting stars in double or triple stellar systems, because of the increased chances of energy fluxes sterilizing the nascent life through sudden heat or cold. And even if complex life did evolve in such planetary systems, it might be difficult for it to survive for any appreciable time.

Our planet was also of suitable size, chemical composition, and distance from the sun to enable life to thrive. An animal-inhabited planet must be a suitable distance from the star it orbits, for this characteristic governs whether the planet can maintain water in a liquid state, surely a prerequisite for animal life as we know it. Most planets are either too close or too far from their respective stars to allow liquid water to exist on the surface, and although many such planets might harbor simple life, complex animal life equivalent to that on Earth cannot long exist without liquid water.

Few Impacts

Another factor clearly implicated in the emergence and maintenance of higher life on Earth is our relatively low asteroid or comet impact rate. The collision of asteroids and comets with a planet can cause mass extinctions, as we have noted. What controls this impact rate? The amount of material left over in a planetary system after formation of the planets influences it: The more comets and asteroids there are in planet-crossing orbits, the higher the impact rate and the greater the chance of mass extinctions due to impact. Yet this may not be the only factor. The types of planets in a system might also affect the impact rate and thus play a large and unappreciated role in the evolution and maintenance of animals. For Earth, there is evidence that the giant planet Jupiter acted as a "comet and asteroid catcher," a gravity sink sweeping the solar system of cosmic garbage that might otherwise collide with Earth. It thus reduced the rate of mass extinction events and so may be a prime reason why higher life was able to form on this planet and then maintain itself. How common are Jupiter-sized planets?

In our solar system, Earth is the only planet (other than Pluto) with a moon of such appreciable size compared to the planet it orbits, and it is the only planet with plate tectonics, which causes continental drift. As we will try to show, both of these attributes may be crucial in the rise and persistence of animal life.

Perhaps even a planet's placement in a particular region of its home galaxy plays a major role. In the star-packed interiors of galaxies, the frequency of supernovae and stellar close encounters may be high enough to preclude the long and stable conditions apparently required for the development of animal life. The outer regions of galaxies may have too low a percentage of the heavy elements necessary to build rocky planets and to fuel the radioactive warmth of planetary interiors. The comet influx rate may even be affected by the nature of the galaxy we inhabit and by our solar system's position in that galaxy. Our sun and its planets move through the Milky Way galaxy, yet our motion is largely within the plane of the galaxy as a whole, and we undergo little movement through the spiral arms. Even the mass of a particular galaxy might affect the odds of complex life evolving, for galactic size correlates with its metal content. Some galaxies, then, might be far more amenable to life's origin and evolution than others. Our star—and our solar system—are anomalous in their high metal content. Perhaps our very galaxy is unusual.

Luck Helps

Finally, it is likely that a planet's *history*, as well as its environmental conditions, plays a part in determining which planets will see life advance to animal stages. How many planets, otherwise perfectly positioned for a history replete with animal life, have been robbed of that potential by happenstance? An asteroid impacting the planet's surface with devastating and life-exterminating consequences. Or a nearby star exploding into a cataclysmic supernova. Or an ice age brought about by a random continental configuration that eliminates animal life through a chance mass extinction. Perhaps chance plays a huge role.

Ever since Polish astronomer Nicholas Copernicus plucked it from the center of the Universe and put it in orbit around the sun, Earth has been periodically trivialized. We have gone from the center of the Universe to a small planet orbiting a small, undistinguished star in an unremarkable region of the Milky Way galaxy—a view now formalized by the so-called Principle of Mediocrity, which holds that we are not the one planet with life but one of many. Various estimates for the number of other intelligent civilizations range from none to 10 trillion.

If it is found to be correct, however, the Rare Earth Hypothesis will reverse that decentering trend. What if the Earth, with its cargo of advanced animals, is virtually unique in this quadrant of the galaxy—the most diverse planet, say, in the nearest 10,000 light-years? What if it is utterly unique: the only planet with animals in this galaxy or even in the visible Universe, a bastion of animals amid a sea of microbe-infested worlds? If that is the case, how much greater the loss the Universe sustains for each species of animal or plant driven to extinction through the careless stewardship of *Homo sapiens?*

The Hubble Space Telescope Should Be Saved

By the National Research Council

The Hubble Space Telescope, in orbit high above Earth, has produced unique images of the universe. The Hubble's success has depended on service visits from manned space shuttles. However, in the wake of the disastrous loss of the shuttle *Columbia* on February 1, 2003, plans for a shuttle mission to the Hubble were canceled. In the following selection a committee at the National Research Council (NRC) summarizes its recommendations for extending the mission of the space telescope. If left unattended, the telescope is likely to fail in 2007, it says. The space telescope's batteries, gyroscopes, and aiming sensors are all liable to break down. The committee considers the possibility of a robotic repair mission, but it rejects this option because it would take until 2010 to launch such an unmanned attempt, and it would not be as effective as a shuttle repair mission using astronauts. In conclusion, the committee recommends that the National Aeronautics and Space Administration reinstate its planned shuttle service mission to keep the Hubble—which the committee calls a "uniquely powerful instrument"—going. The NRC committee was charged by Congress to evaluate options for extending the life of the Hubble Space Telescope. Louis J. Lanzerotti, a distinguished research professor at the New Jersey Institute of Technology, chaired the committee.

The Hubble Space Telescope (HST) was launched from the space shuttle in 1990 and has operated continuously in orbit for the past 14 years. HST was designed to be serviced by

National Research Council, Committee on the Assessment of Options for Extending the Life of the Hubble Space Telescope, *Assessment of Options for Extending the Life of the Hubble Space Telescope: Final Report*. Washington, DC: National Academies Press, 2005.

astronauts, and a series of four shuttle servicing missions from 1993 to 2002 replaced nearly all the key components except the original telescope mirrors and support structure. Three of the four servicing missions added major new instrument observing capabilities. A fifth planned mission, designated SM-4 (servicing mission 4), was intended to replace aging spacecraft batteries, fine-guidance sensors, and gyroscopes and install two new science instruments on the telescope.

Following the loss of the space shuttle Columbia and its crew in February 2003, NASA [National Aeronautics and Space Administration] suspended all shuttle flights until the cause of the accident could be determined and steps taken to reduce the risks of future shuttle flights. In mid-January 2004 NASA decided, on the basis of risk to the astronaut crew, not to pursue the HST SM-4 mission. This cancellation, together with the predicted resulting demise of Hubble in the 2007–2008 time frame, prompted strong objections from scientists and the public alike. NASA continued to investigate options other than a shuttle astronaut mission for extending Hubble's science life and is currently in the early stages of developing an unmanned mission that would attempt to service Hubble robotically. NASA also plans to de-orbit HST by approximately 2013 by means of a robotic spacecraft. . . .

Hubble Components at Risk

The Hubble systems with the greatest likelihood of failing and thus ending or significantly degrading Hubble science operations are the gyroscopes, the batteries, and the fine-guidance sensor (FGS) units. In addition, the HST avionics systems is vulnerable to the aging of the facility.

The telescope uses three gyroscopes to provide precision altitude control. There are currently four functional gyros on HST—three in operation plus one spare. It is likely that the HST system will be reduced to two operating gyros in the latter half of 2006. The HST engineering team is currently working on approaches to sustaining useful, though potentially degraded, astronomical operations with only two gyros, and NASA expects to have that capability by the time it becomes necessary. Eventually, without servicing, the telescope will be reduced to operation with a single gyro in mid to late 2007. The spacecraft can be held in a safe configuration with one or no operating gyros, but science operations will not be possible.

Battery failures are another likely cause of loss of science operations. HST now has six batteries, of which five are necessary for full operations. If battery levels fall too low, the temperature of the structural elements in the Optical Telescope Assembly will fall below permissible levels, causing permanent damage to the facility. Recovery of scientific operations from this state is not possible.

The FGS units (in combination with their electronics subsystems) are used for precision pointing of the observatory. Two operating FGS units are required to support the HST observing program, with a third to supply redundancy. Based on recent test and performance data, one of the three currently operating FGS units is projected to fail sometime between October 2007 and October 2009, and a second is expected to fail sometime between January 2010 and January 2012.

Based on its examination of data and numerous technical reports on Hubble component operations, as well as discussions held with Hubble project personnel, the committee developed the following findings predicated on an estimated SM-4 earliest launch date of July 2006 and a most likely robotic mission launch date of February 2010.

Finding: The projected termination in mid to late 2007 of HST science operations due to gyroscope failure and the projected readiness in early 2010 to execute the planned NASA robotic mission result in a projected 29-month interruption of science operations. No interruption of science operations is projected for a realistically scheduled SM-4 shuttle mission.

Finding: The planned NASA robotic mission is less capable than the previously planned SM-4 shuttle astronaut mission with respect to its responding to unexpected failures and its ability to perform proactive upgrades. Combined with the projected schedule for the two options, the mission risk associated with achieving at least 3 years of successful post-servicing HST science operations is significantly higher for the robotic option, with the respective risk numbers at 3 years being approximately 30 percent for the SM-4 mission and 80 percent for the robotic mission.

Hubble's Contributions

Over its lifetime, HST has been an enormous scientific success, having earned extraordinary scientific and public recognition for its contributions to all areas of astronomy. Hubble is the most powerful space astronomical facility ever built, and it provides wavelength coverage and capabilities that are unmatched by any other optical telescope currently operating or planned.

The four key advantages that Hubble provides over most other optical astronomical facilities are unprecedented angular resolution over a large field, spectral coverage from the visible and the near infrared to the far ultraviolet, access to an extremely dark sky, and highly stable images that enable precision photometry. Hubble's imaging fields of view are also considerable, permitting mapping of extended objects and significant regions of sky. In contrast, ground-based telescopes have a view that is blurred by the atmosphere, and they are completely blind in the ultraviolet and large portions of the near infrared. Hubble can see sharply and clearly at all wavelengths from the far ultraviolet to the near infrared. Hubble images are 5 to 20 times sharper than those obtained with standard ground-based telescopes, in effect bringing the universe that much "closer." Image sharpness and the absence of light pollution in orbit help Hubble to see objects 10 times fainter than even the largest ground-based telescopes. Moreover, Hubble's images are extremely stable, in contrast to those obtained with ground telescopes, whose view is continually distorted by changing atmospheric clarity and turbulence.

Singly, each of these advantages would represent a significant advance for science. Combined, they have made Hubble the most powerful optical astronomical facility in history. Hubble is a general-purpose national observatory that enables unique contributions to and insights concerning most astronomical problems of greatest current interest. Among the most profound contributions of Hubble have been the following:
- Direct observation of the universe as it existed 12 billion years ago,
- Measurements that helped to establish the size and age of the universe,
- Discovery of massive black holes at the center of many galaxies,
- Key evidence that the expansion of the universe is accelerating, which can be explained only by the existence of a fundamentally new type of energy, and therefore new physics, and
- Observation of proto-solar systems in the process of formation.

In addition to its impact on science, Hubble discoveries and images have generated intense public interest. Examples of Hubble data and images that fascinated the public (and scientists) include the big "black eye" left by comet Shoemaker-Levy's direct hit on Jupiter's atmosphere, which alerted the public to the dangers of asteroids impacting Earth; a panoply of jewel-like planetary nebulas that illustrate the ultimate death of our Sun; portraits of planets in the solar system, including auroras on Jupiter and Saturn; and such astronomical spectacles as the "pillars of dust" in the Eagle nebula that appeared on nearly every front page in America and became iconic for Hubble itself. The Hubble Space Telescope has clearly been one of NASA's most noticed science projects, garnering sustained public attention over its entire lifetime.

Maintaining Hubble's Capabilities

The four previous servicing missions to Hubble have added new observing modes and increased existing capabilities, typically by factors of between 10 and 100, since the telescope first flew in 1990. As a result, Hubble now produces more data per unit time than it did originally. The total rate of calibrated data has grown by a factor of 33 since launch. A further increase was expected with the installation of the two new science instruments, the Wide-field Camera 3 (WFC3) and the Cosmic Origins Spectrograph (COS), each of which would provide a greater than 10-fold improvement in scientific efficiency and sensitivity compared with previous instruments. Both of these instruments are already built.

With the installation of WFC3 and COS, and the continued operation enabled by a fifth servicing mission, a broad range of new discoveries would be expected from Hubble. In fact, the committee concluded that Hubble's promise for future discoveries following a fifth servicing mission would be comparable to the telescope's promise when first launched. For example, an important new technique that Hubble would offer for finding planets could enable detection of as many as 1000 new planets in the Milky Way Galaxy in the years after servicing. In addition, a large number of new supernovas could be found for the study of dark energy, reducing uncertainties in its properties by a factor of two. A wealth of data would also be collected to explore the nature of stars in the Milky Way Galaxy and in neighboring galaxies. Hubble is just now beginning to image objects being found by sister NASA missions such as Chandra (an x-ray observatory), Galaxy Evolution Explorer (GALEX; an ultraviolet imager), and Spitzer (an infrared imager and spectrograph), which are currently in orbit. These satellites are relatively wide-field survey telescopes whose goal in part is to detect objects for Hubble follow-up observations. These detailed follow-ups take time because of Hubble's smaller field of view; a large fraction of the scientific benefit of these other satellites will be lost if Hubble's mission is cut short prematurely. And finally, a servicing mission is needed to allow an orderly completion of large, homogeneous data sets such as spectral libraries and imaging surveys of large areas of the Milky Way Galaxy that Hubble is now gathering. These data sets will be archived to serve astronomers for decades to come, given that there are no foreseeable plans to replace Hubble with a telescope of comparable size, wavelength coverage, and high resolution.

The key findings of the committee related to the benefits of future servicing of Hubble are as follows:

While anchored to a remote arm, an astronaut prepares to be elevated to the top of the Hubble Space Telescope for a repair mission.

Finding: The Hubble Space Telescope is a uniquely powerful observing platform in terms of its high angular optical resolution, broad wavelength coverage from the ultraviolet to the near infrared, low sky background, stable images, exquisite precision in flux determination, and significant field of view.

Finding: Astronomical discoveries with Hubble from the solar system to the edge of the universe are among the most significant intellectual achievements of the space science program.

Finding: The scientific power of Hubble has grown enormously as a result of previous servicing missions.

Finding: The growth in the scientific power of Hubble would continue with the installation of the two new instruments, WFC3 and COS, planned for the SM-4 shuttle astronaut mission.

The Risks of Robotic Servicing

Because a robotic servicing mission does not involve risks to the safety of an astronaut crew, the principal concerns are the risk of failure to develop a robotic mission capability in time to service Hubble, and the risk of a mission failure that results in an inability to perform the needed servicing, or worse, critically damages Hubble during the mission. Both schedule risk and mission risk are composed of a large number of factors that were studied in considerable detail by the committee.

Some of the critical components of mission risk include lack of adequate development time to validate the hardware, level of software and system performance required to rendezvous with Hubble, failure to successfully grapple and dock with Hubble, failure to successfully execute the combination of complex autonomous and robotic activities required to actually accomplish HST revitalization and instrument replacement, and the risk of unforeseen Hubble failures prior to mission execution that the robotic mission will not have been designed to repair. One example of a mission risk that concerned the committee is the complicated docking maneuver required for a Hubble robotic servicing, which has never been performed autonomously or teleoperated with time delays. Specifically, the use of the grapple system to autonomously perform close-proximity maneuvers and the final capture of Hubble is a significant challenge and is one of the key technical aspects of a robotic servicing mission that has no precedent in the history of the space program.

The components of schedule risk examined by the committee included the readiness levels of such technologies as the sensors, software and control algorithms, and vision-based closed-loop support for autonomous docking operations, as well as NASA's relevant programmatic and technical expertise, resources, and specific development plans for a robotic servicing mission. From the risk mitigation viewpoint, the committee judged that the planned use of the mature International Space Station robotic arm and robotic operational ground system helps reduce both the schedule risk and the development risk for the robotic mission. In addition, the committee assessed the development schedule for the robotic servicing mission based on its experience with programs of similar complexity and the historical spacecraft development schedule data provided by both NASA and the Aerospace Corporation. The committee's key findings regarding the question of the risk of robotic servicing are as follows:

Finding: The technology required for the proposed HST robotic servicing mission involves a level of complexity, sophistication, and maturity that requires significant development, integration, and demonstration to reach flight readiness and has inherent risks that are inconsistent with the need to service Hubble as soon as possible.

Finding: The Goddard Space Flight Center HST project has a long history of HST shuttle servicing experience but has little experience with autonomous rendezvous and docking or robotic technology development, or with the operations required for the baseline HST robotic servicing mission.

Finding: The proposed HST robotic servicing mission involves a level of complexity that is inconsistent with the current 39-month development schedule and would require an unprecedented improvement in development performance compared with that of space missions of similar complexity. The likelihood of successful development of the HST robotic servicing mission within the baseline 39-month schedule is remote.

Based on extensive analysis, the committee concluded that the very aggressive schedule for development of a viable robotic servicing mission, the commitment to development of individual elements with incomplete systems engineering, the complexity of the mission design, the current low level of technology maturity, the magnitude of the risk-reduction efforts required, and the inability of a robotic servicing mission to respond to unforeseen failures that may well occur on Hubble between now and the mission, together make it highly unlikely that NASA will be able to extend the science life of HST through robotic servicing.

The Risks of Shuttle Servicing

The risks that must be considered in making a decision to service Hubble with the shuttle are the risk to the safety of the crew and the shuttle, as well as the risk of failing to accomplish the servicing objectives. As part of its assessment of safety risk, the committee looked carefully at the findings and recommendations of the Columbia Accident Investigation Board (CAIB) and at NASA's return-to-flight (RTF) requirements. Strong consideration was given to understanding differences in the safety risk factors between shuttle missions to the International Space Station (ISS)—to which NASA still plans to fly 25 to 30 missions—and a shuttle mission to Hubble. Technical considerations examined by the committee included comparisons of on-orbit inspection and repair capabilities at ISS and Hubble, various safe-haven and rescue options, and the likelihood of the shuttle being damaged by micrometeoroid orbital debris (MMOD). With regard to mission risk, the committee considered both the known on-orbit operations required for Hubble servicing and past experience with Hubble shuttle astronaut servicing, including such factors as unforeseen on-orbit contingencies.

The committee developed a large number of findings based on the various analyses cited above. Some of the key findings relevant to the question of the risk of shuttle servicing of HST are as follows:

Finding: Meeting the CAIB and NASA requirements (relative to inspection and repair, safe haven, shuttle rescue, MMOD, and risk to the public) for a shuttle servicing mission to HST is viable.

Finding: The shuttle crew safety risks of a single mission to ISS and a single HST mission are similar and the relative risks are extremely small.

Finding: Previous human servicing missions to HST have successfully carried out unforeseen repairs as well as executing both planned and proactive equipment and science upgrades. HST's current excellent operational status is a product of these past efforts.

Finding: Space shuttle crews, in conjunction with their ground-based mission control teams, have consistently developed innovative procedures and techniques to bring about desired mission success when encountering unplanned for or unexpected contingencies on-orbit.

Finding: The risk in the mission phase of a shuttle HST servicing mission is low.

Comparison of Risks and Benefits

As noted above, the Hubble Space Telescope provides unique capabilities for astronomical research. These capabilities will not be replaced by any existing or currently planned astronomy facility in space or on Earth. Hubble's continuing and extraordinary impact on human understanding of the physical universe has been internationally recognized by scientists and the public alike.

Upgrading Hubble to address the predictable decline in HST component performance over time and thus ensure system reliability requires a timely and successful servicing mission in order to minimize further degradation and prevent a significant gap in science data return. Although it considered other options for extending the life of Hubble, the committee focused on two approaches: robotic servicing and shuttle astronaut servicing.

The need for timely servicing of Hubble imposes difficult requirements on the development of a robotic servicing mission. The very aggressive schedule, the complexity of the mission design, the current low level of technology maturity, and the inability of a robotic servicing mission to respond to unforeseen failures that may well occur on Hubble between now and a servicing mission make it unlikely that the science life of HST will be extended through robotic servicing.

A shuttle astronaut servicing mission is the best option for extending the life of Hubble and preparing the observatory for eventual robotic de-orbit by, for example, attaching targets to Hubble. The committee believes that a shuttle HST servicing mission could occur as early as the seventh shuttle mission following return to flight, at which point critical shuttle missions required for maintaining ISS will have been accomplished. All important systems needed to keep Hubble functioning well through 2011 were included in the original SM-4 shuttle servicing plan. Replacement of batteries and gyros and one FGS is deemed essential. Any spacecraft is subject to unanticipated failures, but if the repairs planned for the SM-4 mission are carried out promptly, there is every prospect that Hubble can operate effectively for another 4 to 5 years after servicing.

The committee finds that the difference between the risk faced by the crew of a single shuttle mission to ISS—already accepted by NASA and the nation—and the risk faced by the crew of a single shuttle servicing mission to HST, is very small. Given the intrinsic value of a serviced Hubble, and the high likelihood of success for a shuttle servicing mission, the committee judges that such a mission is worth the risk.

Recommendations

1. The committee reiterates the recommendation from its interim report that NASA should commit to a servicing mission to the Hubble Space Telescope that accomplishes the objectives of the originally planned SM-4 mission.

2. The committee recommends that NASA pursue a shuttle servicing mission to HST that would accomplish the above stated goal. Strong consideration should be given to flying this mission as early as possible after return to flight.

3. A robotic mission approach should be pursued solely to deorbit Hubble after the period of extended science operations enabled by a shuttle astronaut servicing mission, thus allowing time for the appropriate development of the necessary robotic technology.

The Risks of Saving the Hubble Are Too Great

By Thomas D. Jones

National Aeronautics and Space Administration (NASA) administrator Sean O'Keefe's decision early in 2004 to cancel a space shuttle mission to repair the orbiting Hubble Space Telescope brought howls of protest from around the world. Astronomers, newspapers, the public, and members of Congress were nearly unanimous in criticizing the decision, which experts predicted would doom the orbiting telescope to break down within a few years. In the selection that follows, an astronaut with considerable experience on shuttle missions defends O'Keefe's decision. Thomas D. Jones reminds readers that the cancellation was prompted by the tragic loss of the space shuttle *Columbia*, which broke up on reentry to Earth's atmosphere on February 1, 2003. A shuttle mission to the Hubble Space Telescope would be especially hazardous, Jones argues, because unlike a trip to the International Space Station, if something were to go wrong there would be no refuge for the astronauts. Jones says that as an astronaut he would accept the risks provided the benefits outweighed them. However, he sees little benefit in extending the life of the Hubble when a new space telescope is planned for launch in 2011. By canceling the repair mission, NASA can save money, which can then be spent on the Hubble's replacement, he argues. On the other hand, if a repair mission were sent and another shuttle were lost, the entire space program would be jeopardized, he says. Jones, a graduate of the U.S. Air Force Academy, served as an astronaut on four space shuttle missions, logging many weeks in space.

For nearly 15 years, the Hubble Space Telescope (HST) has been orbiting Earth, cruising through the hard vacuum 350 mi. above the planet. During its long mission, Hubble's passage hasn't created a single sound. Now the prospect of its passing has generated a shock wave of controversy.

NASA's [National Aeronautics and Space Administration] January 16 [2004] decision to cancel further astronaut servicing of the most successful scientific satellite in history generated immediate protests in both the astronomy and political communities. NASA Administrator Sean O'Keefe argued that the cumulative risk of sending a shuttle crew to HST, even for a single life-extension mission, outweighed the scientific benefit. Sen. Barbara Mikulski (D-Md.) quickly asked O'Keefe to reconsider, and he in turn asked Adm. Hal Gehman, who led the [space shuttle] Columbia accident investigation, to review NASA's decision.

Given the Hubble Telescope's scientific success, its role as a cultural icon of space exploration, and its political popularity, arguments about Hubble's fate will continue for some time, especially in light of Gehman's response to the request: "I suggest only a deep and rich study of the entire gain/risk equation can answer the question of whether an extension of the life of the wonderful Hubble telescope is worth the risks involved . . . "

The HST controversy generated a flurry of Internet petitions and op-ed pieces from an array of interested parties, among them the American Physical Society, Mars mission advocate Robert Zubrin, and Rep. Mark Udall (D-Colo.). In early March, Udall introduced a House resolution to have NASA create an independent panel to review O'Keefe's decision. The turmoil surrounding the celebrated but aging telescope guarantees that HST's eventual demise will be just as eventful as its much-heralded (and myopic) debut in orbit in April of 1990.

What the Hubble discussion illustrates is NASA's new sensitivity to the risk of space shuttle operations, and the challenges the nation will face in grappling with the general issue of risk and safety in human space exploration.

Columbia's loss on February 1, 2003, permanently altered the "acceptable risk" equation for human spaceflight. Given the shuttle's inherent limitations, NASA has resolved to minimize the risk to its remaining orbiters.

Why the special concern over a mission to Hubble? The media and many critics have often muddled the technical issues in play, but these can be summarized briefly: NASA would like to restrict future shuttle missions to the orbit of the International Space Station (ISS), both to take advantage of its inspection and repair capabilities, and to have the option of an ISS safe haven in case the orbiter is too badly damaged to return to Earth.

But a shuttle launched to Hubble's 28.5 deg.-inclination orbit cannot carry enough fuel to rendezvous with ISS, whose orbital inclination is 51.6 deg. Instead, a crippled shuttle in Hubble's low-inclination orbit would have to rely on a nail-biting rescue by another orbiter. The "rescue" shuttle would have to be "on standby," ready for rollout and launch within a week or two of the liftoff of the Hubble servicing mission. Two orbiters would thus be dedicated to the next Hubble servicing mission, SM-4, at a time when ISS assembly is putting maximum demand on the three-orbiter fleet. O'Keefe has decided that extending Hubble's scientific life beyond 2007 is not worth putting two shuttles, their crews, and ISS completion on the line.

An Astronaut's Perspective

My personal perception of risk certainly has changed since my first spaceflight in 1994. A 1995 [NASA contractor] SAIC study estimated that a catastrophic accident would occur on about one in every 145 missions. We have, in fact, suffered the loss of two orbiters in 22 years of operations; every time we launch, there's a bit less than a 2% chance that we will lose the vehicle.

I've seen the amazing capability and flexibility of the space shuttle system firsthand, yet it is inextricably tied to the limitations of its design. On two of my shuttle flights, a last-second engine failure and a shattered heat shield tile reminded me of the narrow margins of safety we can expect on flights beyond our atmosphere.

Ten years ago [1994], on a sultry August morning at Kennedy Space Center, I lay strapped in shuttle Endeavour's cabin as the countdown for STS-68 neared zero. With Endeavour's three main engines spooling up to full thrust, we were within 2 sec of liftoff when the general-purpose computers detected an overheating high-pressure-oxygen turbopump on the number three engine, and ordered an emergency shutdown. (A high-pressure turbopump failure is one of the largest contributors to total risk on each flight.)

Had the malfunction occurred a few heartbeats later, after

booster ignition, my crew might have left the pad with one engine down, and been forced to fly a hair-raising Return to Launch Site abort to Kennedy's 3-mi.-long runway. One of the exciting features of an RTLS abort profile is the necessity for the orbiter to fly backward through its exhaust plume at Mach 5. No shuttle crew has ever flown such an emergency approach, and none wants to be the first to try.

When we made it to orbit six weeks later on a new set of engines, we opened Endeavour's payload bay doors to find a disconcerting surprise. Looking aft out the cabin windows, we discovered that one of our thermal protection tiles, atop the starboard orbital maneuvering engine pod, had been partially shattered by debris impact during launch (the culprit was a small tile that broke away from a cabin overhead window).

Fortunately for us, the damaged tile was one that experiences relatively low temperatures during reentry. We surveyed the damage with binoculars, sent Mission Control close-up video for analysis, and then got on with Space Radar Lab 2's demanding Mission to Planet Earth. Looking back on our debris strike, the parallels with Columbia's demise a year ago are chilling. As a crew, we were certainly aware of how lucky we'd been in both our brushes with catastrophe.

Those events of 1994—and of last year—now color my perceptions of the risk in human spaceflight. Sending humans into space is a complex and demanding business. NASA managers must make operational decisions, some with life-and-death implications, knowing that there are valid technical arguments on both sides. Such is the case with O'Keefe's Hubble decision. Though I don't believe a Hubble servicing mission like the cancelled SM-4 is any more risky than a mission to ISS, I can agree with the NASA administrator's decision.

Not Enough Benefit in Fixing Hubble

When I think about astronaut risk, the statistics don't tell the whole story. I don't believe we can predict precisely the likelihood of another shuttle accident based on the 113 missions to date. On every shuttle flight there is a certain irreducible level of inherent risk, driven by everything from main engines to orbital debris to thermal protection system failure—take your pick.

It doesn't matter much to me whether the odds of a catastrophe are 1 in 145 or 1 in 500; what's important is my awareness that the shuttle's design has built-in vulnerabilities that can't be engineered away. If you fly the shuttle (or any spaceship), you face the chance of not coming back. What does matter to me, and to my active astronaut colleagues, is that the benefits of the mission outweigh the very real dangers.

Many astronauts, perhaps a majority, feel that extending Hubble's spectacular run of scientific discovery by three or more years would be worth the risk entailed in a single shuttle mission. As previous Hubble crews have noted, servicing the telescope involves one directly in one of our species' most rewarding pursuits: the effort to understand our universe. Hubble's immense popularity, combined with the public's heightened post-Columbia worries about the shuttle, would make for a dramatic mission, great theater for the press, and a highlight of human endeavor in space.

Would I be willing to climb aboard a shuttle headed for HST? Only if I decided the benefits were worth the risk. Yes, I see the value of extending Hubble's productivity. And I'd get immense satisfaction, as many astronaut colleagues have, from making a personal contribution to one of the greatest scientific enterprises of the past century. But in my view, another servicing mission may harm the nation's prospects for other, more long-lasting exploration efforts. I would not participate in a mission that was at best tangential to humans' ability to explore the solar system first-hand.

Looking Ahead to Future Missions

The president has given NASA a new mission: Extend human presence to the Moon, and sustain a program of human exploration that will eventually reach the asteroids, Mars, and beyond. But he will have a tough problem funding such an ongoing and ambitious exploration program: Only $12 billion has been allocated over the next 10 years, most of that coming from the eventual retirement of the shuttle and the phaseout of major U.S. participation in the ISS. Most agree that won't be nearly enough to accomplish the goals laid out for the next decade.

But NASA has to start somewhere, and the agency appears determined to make choices about how best to make progress toward the Moon and the new vehicles required to get there. The first and most visible of these decisions was the Hubble SM-4 cancellation.

In dropping SM-4, O'Keefe was thinking not only about shuttles and the astronauts who fly them, but also about NASA's new exploration mission. Research aboard ISS into human health and productivity in space is a key enabler for expeditions to the Moon, asteroids, and Mars. Because completing the station's assembly and fulfilling our commitments to international partners will require the three remaining orbiters to fly about 30 missions to the outpost, it becomes essential to preserve the shuttle's remaining capability.

Servicing Hubble would divert at least one orbiter and perhaps a second from ISS missions for about a year (out of the remaining six years allotted for station assembly). Another servicing mission, while effective in extending the telescope's scientific life, would divert scarce resources from the assembly effort. The fact is that deep-space astronomy, a mission not unique to HST, will not directly advance the agency's central human exploration mission. On the other hand, losing an orbiter on a Hubble mission, however unlikely, would halt U.S. human spaceflight in its tracks.

Managing programmatic risk is a mandatory skill in space exploration. Hard choices have been made in the past, often forced by budget as well as technical realities: the cancellation of the final Mercury orbital flight; the shelving of Apollos 18 through 20; the cost-driven compromises in the shuttle's initial design. NASA's current leaders must have the courage to prune off branches that don't promote the growth and vigor of the main stem of the exploration program. Astronauts and managers shouldn't accept risk just to prove they're courageous; often the courageous move is to avoid the risk, especially when the choice is not clear.

On my 1996 Columbia flight, STS-80, mission managers cancelled two spacewalks that my crew had trained a year to execute. But the benefits of our forcing open a jammed airlock hatch weren't worth the risk to our ORFEUS-SPAS space telescope, the "main stem" of our scientific mission. Yes, I was angry. But I couldn't argue with the logic.

Saving Funds for the Next Telescope

Not surprisingly, there is also a financial dimension to the decision. Cancelling SM-4 would free up about $600 million in development and launch costs. Hubble scientific operations cost about another $160 million annually. If Hubble operations end in 2007 (its expected lifetime without another shuttle visit), NASA might then save about $1 billion between now and 2010.

Given early congressional skepticism concerning the funds available for the president's new initiative, those savings are significant. A portion might be applied to Hubble's successor, the James Webb Space Telescope, planned for launch in 2011. The balance will be available to advance the exploration of the Moon, Mars, and asteroids, all rich scientific objectives in their own right.

Critics of O'Keefe's decision tend to overlook the fact that NASA is not abandoning orbital astronomy. Hubble may soldier on productively for another three or four years. The Chandra and Spitzer space observatories, having recently joined Hubble in orbit, are hitting their scientific stride, and the next-generation Webb telescope will carry our infrared vision even deeper into the cosmos. As for NASA's commitment to science, the ambitious human exploration of the Moon, asteroids, and Mars will do even more for our understanding of Earth's history and our own origins than Hubble has done for our knowledge of the cosmos.

Change will be necessary for NASA to explore the solar system, and NASA's O'Keefe seems willing to make the tough calls. That's what he gets paid for, after all. In one sense, those who argue that NASA's first priority should be maintaining Hubble are defending the space status quo. It's an argument for business as usual, and given the post-Columbia landscape, that's an argument for NASA's eventual eclipse.

By making the difficult decision to look beyond Hubble, NASA reduces risk, stays focused on its new mission, and brings closer the day that humanity moves permanently into the universe that Hubble has helped reveal.

A Manned Mission to Mars Would Cost Too Much

By William Tucker

President George W. Bush's proposal for a manned mission to Mars has raised questions about whether the adventure is worth the cost to taxpayers. In the following selection journalist William Tucker presents a skeptical consideration of the value of such a mission. The estimates for initial expenses of the program—some $12 billion over five years—are unlikely to reflect the true costs of the project, Tucker asserts. He notes that manned space projects typically go wildly over budget. The space shuttle, far from breaking even as projected, has to be refurbished after each flight at a cost of about half a billion dollars, Tucker states. The International Space Station has also cost far more than planned, and its main contribution to science has been to show the deleterious effects on the human body of prolonged stays in space. Tucker criticizes the National Aeronautics and Space Administration for conducting business in the most expensive way possible. He argues that the advantages of exploring Mars cannot justify bankrupting the nation. New York–based journalist William Tucker, a college physics major, has written books on environmentalism, crime, and housing. He is a regular columnist for the *American Enterprise.*

President [George W.] Bush's announcement [on January 14, 2004] of a 280-million-mile manned space flight to Mars caught everyone by surprise. Space enthusiasts, arguing for years that NASA [National Aeronautics and Space Administration] had lost its way, were electrified. "We've been stuck in low

William Tucker, "The Sober Realities of Manned Space Flight: With Feet Planted Firmly on the Ground, a Look at the Sky-High Costs and Mixed Results at NASA," *The American Enterprise,* December 2004. Copyright © 2004 by the American Enterprise Institute for Public Policy Research. Reproduced by permission.

Earth orbit for decades," said Louis Friedman, executive director of the Planetary Society. "The goal should be exploration." NASA, still recovering from the recent Space Shuttle catastrophe,[1] was eager to rededicate itself. A quick NASA calculation, however, revealed that the Mars effort would cost nearly $500 billion over 30 years. With funding tight and trouble brewing in the Middle East, Congress decided to ignore the project. After very little debate, the proposal dropped from sight.

That's what happened in 1989 when the elder President Bush proposed a manned flight to Mars by 2019. In 2004, President Bush the younger returned to his father's unfinished business and called for his own revival of space exploration, which would have astronauts getting back to the moon in 2020 and on to Mars thereafter. Only a year before, NASA director Sean O'Keefe was calling such an idea a "Hail Mary pass" and urging the public to be satisfied with slow, steady exploration by robotic probes. But the spirit of adventure seemed to be reasserting itself. As the President put it at a memorial service for the Columbia astronauts last year [2003], "This cause of exploration and discovery is not an option we chose. It is a desire written in the human heart."

Dubious Cost Estimates

The $500 billion price tag has been scaled down considerably. Building on existing research, NASA will begin with $12 billion in spending over the next five years. Among most space veterans, however, these initial estimates are treated as a joke. "The Space Shuttle was originally supposed to break even and fly every two weeks," reminds Greg Klerkx, whose book *Lost in Space* is a critique of NASA. "It ended up costing $500 million per launch, and flying four or five times a year. When President Ronald Reagan first proposed the International Space Station, it was scheduled to be finished in eight years and cost $9 billion. Now it's over $70 billion and still isn't scheduled for completion until 2010.

"This is the picture of a federal agency immune to the competitive influences of the private sector. You see this little metal loop? It's called a carabineer," California space entrepreneur Rick Tumlinson told a Senate hearing right after the President's announcement in January. "You could go to any sporting goods shop and

---

1. In 2003 the manned space shuttle *Columbia* broke up upon reentry.

buy it for $20. Yet NASA pays over $1,000 for the same object because of its procurement methods. It's the 'not-invented-here'
mentality and distrust of the private sector that makes the cost of
these projects so astronomical."

So will the Mars expedition be different? "The obstacles in getting to Mars are going to be bureaucratic, not technological," says
Howard McCurdy, a space program expert at American University.
"The NASA that got to the moon in 1969 was a totally different
animal from the NASA that got the job in 1961. The current NASA
may have to undergo the same kind of transformation."

NASA has always been a mix of science and show business.
"The moon expedition was basically an episode of the Cold War,"
says Dr. Eligar Sadeh, assistant professor at the Odegard School
of Aerospace Sciences. "Planting the flag, putting footprints on
the moon—that was done to prove we had a better system than the
Soviet Union. Since then, NASA has never been able to refocus
its mission."

[Astronaut] Neil Armstrong's "One giant step for mankind" defined a generation. Yet few people remember that after 1972 the
last three Apollo missions were canceled because the public was
losing interest. There are only so many times you can hit a golf
ball on the moon. "There's a need for heroics," says Supriya
Chakrabarti, director of the Boston University Center for Space
Physics. "It's hard for the public to get excited about astronomers
looking at squiggly lines on their computer screens."

The Space Shuttle

Once the moon had been reached, Mars seemed the next obvious
destination. Instead, President [Richard] Nixon scaled back the
program. At the time, NASA was absorbing an unsustainably large
4 percent of the federal budget. The next step became the Space
Shuttle—a workhorse that seemed eminently practical at the time
but has turned out to be an expensive, clumsy albatross. With the
exception of a few spectacular missions such as the repair of the
Hubble Space Telescope, "the plain fact is that the shuttle has very
little to do other than ferrying cargo to the International Space Station," sighs Louis Friedman of the Planetary Society.

"I vividly remember President Ronald Reagan going down to
Texas after the Challenger disaster and memorializing the astronauts as heroes going further and faster into the unknown" says

Friedman. "Of course they were doing no such thing. They were simply launching a communications satellite and carrying on the teacher-in-space show for schoolchildren—both rather mundane tasks. But by once again evoking the exploration theme, Reagan saved the space program."

Each Space Shuttle (there are now three) must be virtually reconstructed after each flight. The process takes two months and 20,000 people. Some of the parts are so outdated that engineers troll eBay for replacements. "What we save on re-use, we throw away on maintenance of the aging fleet," says Alex Roland, professor of military history at Duke and former historian of NASA. The Shuttles are scheduled to be retired for good in 2010. The Shuttle's main task, launching communications satellites, could be performed just as well by expendable launch rockets. Fearing the program would have little business if this were allowed, Congress mandated that all government satellites be carried aloft by a Shuttle. "This very nearly ended the production of launch vehicles in this country," says Klerkx. "The powerful Saturn rocket, which carried Apollo to the moon, was abandoned after 1970.". . .

Costly Space Station

Today's International Space Station (ISS) was originally proposed as "Space Station Freedom" in 1984 by the Reagan administration. The orbiting outpost was conceived as an extraterrestrial mini-city along the lines suggested by Wernher von Braun—in a series of *Collier's* articles in the 1950s—and MIT [Massachusetts Institute of Technology] visionary Gerard O'Neill in his 1977 book, *The High Frontier*, which suggested 20 million people could live in space. By 1992, however, after spending $11 billion, Space Station Freedom was still on the drawing board. Sensing a quagmire, the Clinton administration brought the Russians in on the project, both to tap their experience and to prevent newly unemployed Russian military scientists from peddling their expertise to rogue nations.

Although scaled down considerably, the freshly christened International Space Station was to include platforms for launching flights to the moon and Mars, and for housing dozens of astronauts carrying on experiments in drug manufacturing, protein crystallization, and molding perfect ball bearings. All this would help the ISS pay for itself, and in June 1993, the House authorized another $13 billion by one vote. But since the first stage was lifted into orbit by the Russians in 1995, costs have soared while ambitions faded. Assembly platforms for launching moon rockets have vanished. When completed, the ISS will hold six astronauts. The two in residence now spend 85 percent of their time on construction and maintenance. In essence, the U.S. is spending billions so that two astronauts can build a space shed.

The only experiments the astronauts are performing are on themselves, measuring the long term effects of life in zero gravity. The news has not been good. Muscles atrophy quickly and—for reasons yet unknown—the human body does not manufacture bone tissue in space. Russian cosmonauts return from the ISS virtually helpless for the first few days. This doesn't bode well for sending astronauts on an 18 month journey to Mars. Artificial gravity could alleviate the problem, but it can't be tested on the Space Station since it requires a large revolving body like the Jupiter probe in [novelist] Arthur Clarke's *2001*.

Meanwhile, the ISS has sucked up funds from nearly every other NASA project. Russia has been unable to pay its share, so the U.S. quietly picks up the tab. "It's hard to pin down because of NASA's accounting, but it looks like we've spent about $80 billion thus far," says Dr. Sadeh, author of 2003's *Space Politics and Policy*. "By the time it's finished it will probably cost $150 billion." Yet the Bush plan calls for continued expansion, adding European and Japanese modules over the next few years. "Right now the argument for completing the ISS is that we have to fulfill our international obligations," adds Klerkx. "Otherwise it would probably be put in mothballs."

Success of Unmanned Efforts

Ironically, while NASA's manned efforts have been of dubious value, its unmanned probes have been hugely successful. Many scientists are calling the last 30 years a "golden age of astronomy" thanks to their discoveries. The Viking, which mapped Mars in the 1970s; the Voyager trips to Jupiter, Saturn, Uranus, and Neptune in the late 1970s and early 1980s; the Galileo probe of the 1990s, which explored Jupiter's moons and discovered an apparent salty ocean beneath the icy crust of Europa; the Cassini-Huygens mission which this year reached Saturn and its moon Titan; the Spirit and Opportunity rovers investigating Mars at this moment—all have expanded our knowledge to a spectacular degree.

To be sure, there have been embarrassments. The Mars Polar Lander failed in 1999 because someone forgot to convert English measurements to metric. Genesis, which probed the sun's environment, recently crashed in the Utah desert when its parachute failed to open. And of course the Hubble Telescope turned out to be myopic—a manufacturing error 1/50th the width of a human hair left the original telescope out of focus.

But that was mostly repaired when Shuttle astronauts fitted it with a "contact lens," and NASA's "eye in the sky" has since returned a decade's worth of breathtaking images and fresh knowledge. Hubble has mapped the heavens, observed gravitational lenses that confirm [Albert] Einstein's theory of general relativity, discovered galaxies that go back almost to the beginning of time, and made a convincing case for the "Dark Energy" that appears to be driving the expansion of the universe.

Abandoning the Hubble Telescope

Yet a week after Bush's Mars proposal, O'Keefe announced that NASA would abandon Hubble. New safety precautions require that the Shuttle be able to dock at the International Space Station in case of any emergency. Since the telescope orbits far from the ISS, any mission bound for Hubble for one of its regularly required maintenance visits would be unable to reach the Space Station if it ran into trouble.

"Hubble's last gyroscopes are due to fail within the next three to four years," says Dr. Sadeh. "After that it can't be positioned for observations." Pressure from the scientific community and Congress has forced O'Keefe to explore robotic repair, but many remain skeptical. "I wouldn't be surprised if a couple of astronauts volunteer to risk one last mission," says Dr. Sadeh.

So NASA's manned programs had just about come to the end of the line when President Bush announced his plan to visit Mars. The message sent a jolt of electricity through the ranks of space enthusiasts, who have been pushing for the new goal for more than a decade. "This was long overdue," says Friedman.

But the Mars mission raises an immediate question—will a reinvigorated NASA finally open space exploration to competition and the private sector? Or will it simply use the Mars mandate to pile cost upon cost in the same old way? "When Sir Ed-

mund Hillary first climbed Mount Everest, he had so many porters
he needed ten more just to carry the money to pay the others," says
McCurdy, who analyzes the space program in his book *Faster,
Cheaper, Better.* "Now two people can climb Mount Everest by
themselves. We need the same approach to Mars. You've got to
pare down. Closing down the Space Station and the Shuttle would
be a start."

Dozens of companies have been hot for chances to launch pas-
sengers and cargo into space. Visionaries have suggested that cash
prizes could encourage the necessary innovation and competition.
Tumlinson has proposed that NASA offer a prize of $50 million
to any private company that can map the south pole of Mars,
where a landing is likely to take place. "People don't remember
that when Charles Lindbergh crossed the Atlantic, he was com-
peting for the $25,000 Orteig prize," he says.

In 1995 Greg Maryniak, a St. Louis trial lawyer, founded the
Ansari X-Prize, offering $10 million to the first group to launch a
passenger-carrying spaceship 100 kilometers above the Earth, return it to the ground, and launch it again within two weeks. On
October 4 [2004], the Mojave Aerospace Ventures Team claimed
the prize when pilot Brian Binnie took SpaceShipOne 368,000
feet above the Earth for the second time in a week. Almost simultaneously, British entrepreneur Richard Branson announced he
would soon be carrying space tourists above the atmosphere for
$200,000 a seat.

"The mission of NASA should be the same as the mission of
the old National Advisory Commission for Aeronautics in the
1920s," says Klerkx. "NACA continued to do research but turned
the building and flying of airplanes over to the private sector." In
1990, an obscure engineer at Martin Marietta named Robert
Zubrin caused a sensation by proposing Mars Direct, a way of getting to the Red Planet without stopping at either the ISS or the
moon. Zubrin's plan would send a robotic Earth Return Vehicle
two years ahead of time to mine the planet for the methane needed
for the return trip.

"We're much better prepared to get to Mars now than we were
to get to the moon in 1961," says Zubrin, who has since founded
the Mars Society and written *The Case for Mars* (1996). "The
whole mission could be accomplished in a decade, rather than 20
years, as NASA is proposing." Yet NASA's not-invented-here policy makes such outside suggestions unlikely.

One of the keys to NASA's plan is the development of a new system for generating electricity from nuclear radiation. In 2002, O'Keefe launched the Prometheus Project, a $1 billion effort to develop an improved nuclear propulsion system. Conventional reactors use radioactive heat to drive steam turbines, but such boilers are too heavy for space.

Another problem: Astronauts making the 560-million-mile round trip will be exposed to huge doses of cosmic radiation. The only protection is a shield made of the heaviest elements—which would weigh down the ship. All this raises an obvious question: Why send people to Mars at all? Couldn't we just rely on improved robotics? "The risk to human life is obviously great," says Friedman. "But the payoff for sending astronauts is huge. A human being exploring the geology of Mars could do in one day what a robot can do in a month."

Doubts About Space Colonization

This leads to an even bigger question: Why go in the first place? Other than planting the flag and establishing our hegemony over the solar system, where is the lasting purpose in such an effort? Viewers from Manila to Moscow would be glued to their TV screens, and nations of the world might feel united (except for Islamic radicals, who would be scheming on how to blow the thing up). But the Olympics serve much the same purpose.

The justifications for human space exploration are generally stated as these: adventure and the stimulation of pushing into the unknown; making new worlds habitable for future generations; scientific discovery and unraveling the mystery of who we are. These are no small things. European cultures were vastly invigorated during the Renaissance by the discovery of the New World. Columbus, Vasco da Gama, and Magellan became heroes who defined the West for centuries. Cultures that pioneer seem to thrive, while those that stop pioneering often fester and degenerate. "Human beings either look out or they look down," says Friedman.

Space enthusiasts have always been big fans of Frederick Jackson Turner's "Frontier Thesis," which says that the American character was formed by the constant exploration of new horizons. "Lots of people stayed home in Europe during the settling of the New World," says McCurdy. "Maybe there's been some selection in our genes."

Whether it is our destiny to colonize the solar system, however, is an entirely different matter. The moon is not a virgin continent waiting to be inhabited, but a barren oxygen-less desert that will have to be claimed inch by inch. At best, it will require the construction of huge, closed-in Earth-like environments that would have to be provisioned continuously. Space enthusiasts often suggest other planets could be "terraformed"—transformed into an Earth-like environment—in case humans have to escape if the Earth should be destroyed by an asteroid. This may make good science fiction but can hardly serve as a goal for NASA policy.

Cost of Curiosity

Finally, there is the matter of scientific investigation. The great question that hovers over all space exploration is the oldest philosophical quest: How did we get here? Are we alone? Are other planets inhabited? Is life a miraculous one-time occurrence or a normal process in the evolution of the universe? These great questions have puzzled humanity since Empedocles asked, "Why is there something instead of nothing?"

As Isaac Newton once said in describing his own scientific motivations, we are sitting on the seashore amused by a few pebbles while whole vast oceans of the unknown lie before us. No one would suggest we should not pursue these mysteries. The question is whether we have to squander vast amounts of money to do it.

Space travel will come eventually. It was more than a century before Columbus's discoveries led to any attempts to colonize the New World. Our timetable may not be all that different. Yet we should also be chastened by the example of sixteenth-century Spain, which bankrupted itself in pursuing the Age of Discovery.

Now as then, "privateering" may be a solution. It is certainly better than shoveling money at an entrenched bureaucracy.

A Manned Mission to Mars Is Vital to the Nation

By Robert Zubrin

In the following selection Mars Society founder Robert Zubrin argues that it is in the nation's vital interests to send humans to Mars within a decade. The scientific payoff could be immense, he argues, because Mars exploration is likely to help answer the question of whether life arises naturally where water and other chemical precursors are present. A manned mission to Mars could also inspire a new generation to enter scientific professions. Most important, however, the venture to Mars could be the first step toward an extraterrestrial human colony. Mars can provide the necessities for life, he argues. Zubrin describes a method for preparing the way for human explorers by sending a robotic mission in advance. The unmanned craft, he says, can begin manufacturing rocket fuel and oxygen from locally available materials on the planet. Then a human crew would follow two years later. Once the first mission is established, crews can change every year or two, providing continuous habitation and exploration. Such heroic enterprises, he argues, are vital to the continuation of Western civilization. An aeronautical engineer by training, Robert Zubrin is founder and president of the Mars Society, which advocates for space exploration. He is also the author of *The Case for Mars.*

I t is not enough that NASA's [National Aeronautics and Space Administration] human exploration efforts "have a goal." The goal selected needs to be the *right* goal, chosen not because various people are comfortable with it, but because there is a real reason to do it. We don't need a nebulous, futuristic "vision" that can be used to justify random expenditures on various fascinating technologies that might plausibly prove of interest at some time in the future when NASA actually has a plan. Nor do we need strategic plans that are generated for the purpose of making use of such constituency-based technology programs. Rather, the program needs to be organized so that it is the goal that actually drives the efforts of the space agency. In such a destination-driven operation, NASA is forced to develop the most practical plan to reach the objective, and on that basis, select for development those technologies required to implement the plan. Reason chooses the goal. The goal compels the plan. The plan selects the technologies.

So what should the goal of human exploration be? In my view, the answer is straightforward: Humans to Mars within a decade. Why Mars? Because of all the planetary destinations currently within reach, Mars offers the most—scientifically, socially, and in terms of what it portends for the human future.

The Search for Life

In scientific terms, Mars is critical, because it is the Rosetta Stone for helping us understand the position of life in the universe. Images of Mars taken from orbit show that the planet had liquid water flowing on its surface for a period of a billion years during its early history, a duration five times as long as it took life to appear on Earth after there was liquid water here. So if the theory is correct that life is a naturally occurring phenomenon, emergent from chemical complexification wherever there is liquid water, a temperate climate, sufficient minerals, and enough time, then life should have appeared on Mars. If we go to Mars and find fossils of past life on its surface, we will have good reason to believe that we are not alone in the universe. If we send human explorers, who can erect drilling rigs which can reach underground water where Martian life may yet persist, we will be able to examine it. By doing so, we can determine whether life on Earth is the pattern for all life everywhere, or alternatively, whether we are simply one esoteric example of a far vaster and more interesting tapestry. These things are truly worth finding out.

In terms of its social value, Mars is the bracing positive challenge that our society needs. Nations, like people, thrive on challenge and decay without it. The challenge of a humans-to-Mars program would be an invitation to adventure to every young person in the country, sending out the powerful clarion call: "Learn your science and you can become part of pioneering a new world." This effect cannot be matched by just returning to the Moon, both because a Moon program offers no comparable potential discoveries and also because today's youth cannot be inspired in anything like the same degree by the challenge to duplicate feats accomplished by their grandparents' generation.

Inspiring Youth to Enter Science

There will be over a hundred million kids in our nation's schools over the next ten years. If a Mars program were to inspire just an extra one percent of them to pursue a scientific education, the net result would be one million more scientists, engineers, inventors, and medical researchers, making technological innovations that create new industries, find new cures, strengthen national defense, and generally increase national income to an extent that utterly dwarfs the expenditures of the Mars program.

But the most important reason to go to Mars is the doorway it opens to the future. Uniquely among the extraterrestrial bodies of the inner solar system, Mars is endowed with all the resources needed to support not only life but the development of a technological civilization. In contrast to the comparative desert of the Moon, Mars possesses oceans of water frozen into its soil as ice and permafrost, as well as vast quantities of carbon, nitrogen, hydrogen, and oxygen, all in forms readily accessible to those clever enough to use them. These four elements are the basic stuff not only of food and water, but of plastics, wood, paper, clothing— and most importantly, rocket fuel.

In addition, Mars has experienced the same sorts of volcanic and hydrologic processes that produced a multitude of mineral ores on Earth. Virtually every element of significant interest to industry is known to exist on the Red Planet. While no liquid water exists on the surface, below ground is a different matter, and there is every reason to believe that underground heat sources could be maintaining hot liquid reservoirs beneath the Martian surface today. Such hydrothermal reservoirs may be refuges in which survivors of ancient Martian life continue to persist; they would also represent oases providing abundant water supplies and geothermal power to future human settlers. With its 24-hour day-night cycle and an atmosphere thick enough to shield its surface against

solar flares, Mars is the only extraterrestrial planet that will readily allow large scale greenhouses lit by natural sunlight. In other
words: Mars can be settled. In establishing our first foothold on
Mars, we will begin humanity's career as a multi-planet species.

   Mars is where the science is, Mars is where the challenge is, and
Mars is where the future is. That's why Mars must be our goal.

The Spirit of Apollo

The faint of heart may say that human exploration of Mars is too
ambitious a feat to select as our near-term goal, but that is the view
of the faint of heart. From the technological point of view, we're
ready. Despite the greater distance to Mars, we are *much* better
prepared today to send humans to Mars than we were to launch
humans to the Moon in 1961 when John F. Kennedy challenged
the nation to achieve that goal—and we got there eight years later.
Given the will, we could have our first teams on Mars within a
decade.

   The key to success is rejecting the policy of continued stagna-
tion represented by senile Shuttle Mode thinking, and returning to
the destination-driven Apollo Mode of planned operation that allowed the space agency to perform so brilliantly during its youth.
In addition, we must take a lesson from our own pioneer past and
adopt a "travel light and live off the land" mission strategy similar to that which has well-served terrestrial explorers for centuries.
The plan to explore the Red Planet in this way is known as Mars
Direct. Here's how it could be accomplished.

   At an early launch opportunity—for example 2014—a single
heavy lift booster with a capability equal to that of the Saturn V
used during the Apollo program is launched off Cape Canaveral
and uses its upper stage to throw a 40-tonne unmanned payload
onto a trajectory to Mars. (A "tonne" is one metric ton.) Arriving
at Mars eight months later, the spacecraft uses friction between its
aeroshield and the Martian atmosphere to brake itself into orbit
around the planet, and then lands with the help of a parachute.
This is the Earth Return Vehicle (ERV). It flies out to Mars with
its two methane/oxygen driven rocket propulsion stages unfueled.
It also carries six tonnes of liquid hydrogen, a 100-kilowatt nuclear reactor mounted in the back of a methane/oxygen driven
light truck, a small set of compressors and an automated chemical processing unit, and a few small scientific rovers.

Making Its Own Fuel

As soon as the craft lands successfully, the truck is telerobotically driven a few hundred meters away from the site, and the reactor is deployed to provide power to the compressors and chemical processing unit. The ERV will then start a ten-month process of fueling itself by combining the hydrogen brought from Earth with the carbon dioxide in the Martian atmosphere. The end result is a total of 108 tonnes of methane/oxygen rocket propellant. Ninety-six tonnes of the propellant will be used to fuel the ERV, while 12 tonnes will be available to support the use of high powered, chemically fueled, long range ground vehicles. Large additional stockpiles of oxygen can also be produced, both for breathing and for turning into water by combination with hydrogen brought from Earth. Since water is 89 percent oxygen (by weight), and since the larger part of most foodstuffs is water, this greatly reduces the amount of life support consumables that need to be hauled from Earth.

With the propellant production successfully completed, in 2016 two more boosters lift off from Cape Canaveral and throw their 40-tonne payloads towards Mars. One of the payloads is an unmanned fuel-factory/ERV just like the one launched in 2014; the other is a habitation module carrying a small crew, a mixture of whole food and dehydrated provisions sufficient for three years, and a pressurized methane/oxygen powered ground rover.

Upon arrival, the manned craft lands at the 2014 landing site where a fully fueled ERV and beaconed landing site await it. With the help of such navigational aids, the crew should be able to land right on the spot; but if the landing is off course by tens or even hundreds of kilometers, the crew can still achieve the surface rendezvous by driving over in their rover. If they are off by thousands of kilometers, the second ERV provides a backup.

Assuming the crew lands and rendezvous as planned at site number one, the second ERV will land several hundred kilometers away to start making propellant for the 2018 mission, which in turn will fly out with an additional ERV to open up Mars landing site number three. Thus, every other year two heavy lift boosters are launched, one to land a crew, and the other to prepare a site for the next mission, for an average launch rate of just one booster per year to pursue a continuing program of Mars exploration. Since in a normal year we can launch about six shuttle stacks, this would only represent about 16 percent of the U.S. heavy-lift capability, and would clearly be affordable. In effect, this "live off the land" approach removes the manned Mars mission from the realm of mega-spacecraft fantasy and reduces it in practice to a task of comparable difficulty to that faced in launching the Apollo missions to the Moon.

Wide-Ranging Exploration

The crew will stay on the surface for 1.5 years, taking advantage of the mobility afforded by the high-powered chemically-driven ground vehicles to accomplish a great deal of surface exploration. With a 12-tonne surface fuel stockpile, they have the capability for over 24,000 kilometers worth of traverse before they leave, giving them the kind of mobility necessary to conduct a serious search for evidence of past or present life on Mars. Since no one has been left in orbit, the entire crew will have available to them the natural gravity and protection against cosmic rays and solar radiation afforded by the Martian environment, and thus there will not be the strong pressure for a quick return to Earth that plagues other Mars mission plans based upon orbiting mother-ships with small landing parties. At the conclusion of their stay, the crew returns to Earth in a direct flight from the Martian surface in the ERV. As the series of missions progresses, a string of small bases is left behind on the Martian surface, opening up broad stretches of territory to human cognizance.

In essence, by taking advantage of the most obvious local resource available on Mars—its atmosphere—the plan allows us to accomplish a manned Mars mission with what amounts to a lunar-class transportation system. By eliminating any requirement to introduce a new order of technology and complexity of operations beyond those needed for lunar transportation to accomplish piloted Mars missions, the plan can reduce costs by an order of magnitude and advance the schedule for the human exploration of Mars by a generation. . . .

Exploring Mars requires no miraculous new technologies, no orbiting spaceports, and no gigantic interplanetary space cruisers. We don't need to spend the next thirty years with a space program mired in impotence, spending large sums of money on random projects and taking occasional casualties while the missions to nowhere are flown over and over again, and while professional technologists dawdle endlessly in their sandboxes without producing the needed flight hardware. We simply need to choose our destination, and with the same combination of vision, practical thinking, and passionate resolve that served us so well during Apollo, do what is required to get us there.

We can establish our first small outpost on Mars within a decade. We, and not some future generation, can have the honor of being the first pioneers of this new world for humanity. All that is needed is present day technology, some nineteenth-century industrial chemistry, a solid dose of common sense, and a little bit of moxie.

The Importance of Science

So we can do it, and it should be done, but why should we be the ones to do it? Why, at a time like this, with the nation at war, with new menaces threatening to appear in various corners of the globe, and our allies drifting away, should the United States government expend serious resources on such a visionary enterprise? In my view, such considerations simply make the matter all the more urgent.

While I would not deny the necessity of military action in certain circumstances, in the long run civilizations are built by ideas, not swords. The central idea at the core of Western civilization is that there is an inherent facility in the individual human mind to recognize right from wrong and truth from untruth. This idea is the source of our notions of conscience and science, terms which, not coincidentally, share a common root.

Both our radical fundamentalist and our totalitarian enemies deny these concepts. They deny the validity of the individual conscience, and they deny the necessity of human liberty, and indeed, consider it intolerable. For them, conscience, reason, and free will must be crushed so that humans will submit to arbitrary and cruel authority.

Against this foe, science is our strongest weapon, not simply because it produces useful devices and medical cures, but because it demonstrates the value of a civilization based upon the use of reason. There was a time when we celebrated the divine nature of the human spirit by building Gothic cathedrals. Today we build space telescopes. Science is our society's sacred enterprise; through it we assert the fundamental dignity of man. And because it ventures into the cosmic realm of ultimate truth, space exploration is the very banner of science.

If the United States is to lead the West, it must not only carry its sword, but the banner of its most sacred cause. And that cause is the freedom to explore on the wings of human reason. The French may sneer, with some cause, at our fast food restaurants and TV sitcoms, but the Hubble Space Telescope can inspire nothing but admiration, or even awe, in anyone who is alive above the neck. A human Mars exploration program would be a statement about ourselves, a reaffirmation that we remain a nation of pioneers, the vanguard of humanity, devoted to the deepest values of Western civilization. But even more, it would be a declaration of the power of reason, courage, and freedom writ clear across the heavens.

Now, more than ever, we need to make those statements. Now, more than ever, we need to sign that declaration—in handwriting large enough that no one will need spectacles to read it.

CHRONOLOGY

ca. 3000 B.C.

Stonehenge, a temple whose monoliths seem to align with the sun at the solstice, is constructed by a people whose history has been lost.

2296

Chinese astronomers make the first recorded observation of a comet.

ca. 2000

Ancient Egyptians create a calendar based on astronomical observations and flood seasons.

586

Thales of Miletus predicts a solar eclipse.

ca. 350

Aristotle, pointing to the shadow on the moon's surface during a lunar eclipse, argues that the earth is a sphere.

ca. 270

Aristarchus of Samos estimates the distance to the sun and proposes that Earth goes around it.

ca. 250

Eratosthenes estimates the size of the earth by comparing the times that the sun illuminates the bottom of wells in different locations; he also identifies the earth's tilt and makes rather less accurate estimates of the distance to the sun and moon.

240

Chinese astronomers record an observation of Halley's comet.

ca. 130
Hipparchus of Rhodes discovers precession of the equinoxes, estimates the distance to the moon, and produces the first accurate star chart.

ca. A.D. 140
In Alexandria, Claudius Ptolemy writes the *Almagest,* an Earth-centered treatise on astronomy that provides a way to predict the movement of planets.

ca. 300
Mayan astronomers begin to keep careful observations of the sun, moon, and planets, leading to a remarkably accurate calendar.

ca. 820
Caliph Al-Ma'mun establishes an academy of science called the "House of Wisdom" in Baghdad, in present-day Iraq, which becomes a center for the study of astronomy.

964
Persian astronomer 'Abd al-Rahman al-Sufi records the existence of the Andromeda galaxy, calling it "little cloud."

1054
Chinese astronomers observe a supernova in the vicinity of the constellation Taurus. It is now known as the Crab Nebula.

1120
A short-lived observatory is built in Cairo, Egypt, the first of several in the Islamic Empire.

1259
The Maragha observatory is built in present-day Iran for the famous Persian astronomer Nasir al-Din al-Tusi.

1543
Nicolaus Copernicus publishes his heliocentric theory of the solar system.

1572

Tycho Brahe discovers a supernova in the constellation of Cassiopeia; he becomes Denmark's first royal astronomer.

1577

Brahe records the transit of a huge comet.

1600

Astronomical philosopher Giordano Bruno, who produced early concepts of relativity and infinity, is burned at the stake in Rome for heresy.

1607

Johannes Kepler, successor to Brahe, observes Halley's comet.

1609

Kepler publishes the first two of his three laws of planetary motion.

1610

Galileo Galilei, having improved the telescope, observes the moons of Jupiter, Saturn, and the phases of Venus as it orbits the sun. Mathematician and astronomer Thomas Harriot of England makes the first observations of sunspots, which eventually allow him to compute the rate of the sun's rotation (which Galileo also does).

1632

Galileo's book *Dialogue Concerning the Two Chief World Systems* is published. Within months Pope Urban VIII bans it and orders its author to appear before the Inquisition.

1633

Under threat of torture and death, Galileo recants the position he took in his book in favor of the Copernican view of the solar system. The Inquisition sentences him to house arrest, where he remains until his death in 1642.

1655

Christian Huygens of the Netherlands discovers Titan, one of Saturn's moons.

1666

Giovanni Cassini of Italy discovers Martian polar ice caps and divisions in the rings of Saturn.

1687

At Cambridge University Isaac Newton publishes his theory of universal gravitation in his masterwork *Principia Mathematica,* regarded as the keystone of modern physics and astronomy.

1705

Edmond Halley of England correctly computes the orbit of the comet that bears his name, allowing him to predict its return in 1758.

1781

British astronomer William Herschel discovers the planet Uranus. Charles Messier of France catalogs fuzzy objects in the sky (now known as Messier Objects), which are later discovered to include galaxies, nebula, star clusters, and comets. The letter *M* continues to be used in the official names of galaxies, clusters, and nebula.

1784

Herschel observes the seasonal variations in the Martian polar ice caps.

1846

Prompted by the mathematical extrapolations of French astronomer Urbain Leverrier, German astronomer Johann Galle discovers the planet Neptune.

1860

Gustav Kirchoff and Robert Bunsen discover that each element has its own distinct set of spectral lines and use this fact to explain the solar dark lines.

1872

American astronomer Henry Draper makes the first spectrum photograph of the distant star Vega. Studying the spectral radiation and absorption lines in light allows astronomers to determine what elements are present in the star.

1905

Albert Einstein introduces the special theory of relativity, which leads to a revolution in the understanding of motions of celestial bodies.

1911

Ejnar Hertzsprung and Henry Russell introduce a diagram that shows how the color and the brightness of typical stars are related.

1915

Einstein publishes the general theory of relativity, which predicts that the gravity of a massive object will warp the space-time around it.

1919

British astronomer Arthur Eddington vindicates Einstein's general theory of relativity by photographing stars close to the sun during a solar eclipse and then comparing their position in the sky six months later. His findings show that, as predicted, the sun's gravity bent the starlight, making the stars appear to shift.

1923

Edwin Hubble proves that other galaxies lie outside our own by discovering Cepheids (predictably variable stars) in the Andromeda galaxy, comparable to those in the Milky Way.

1929

Robert Goddard launches the first instrument-bearing rocket. Hubble, working from observations of numerous galaxies, publishes a preliminary finding that all galaxies are receding from one another.

1930

American astronomer Clyde Tombaugh discovers Pluto.

1931

Karl Jansky of Bell Labs discovers cosmic radio waves, opening the way to radio astronomy.

1947

A major radio telescope, later named for its developer, Bernard Lovell, goes online at Jodrell Bank, in northern England.

1948

George Gamow, Ralph Alpher, and Robert Herman predict that the big bang will have produced cosmic microwave background radiation. (Their paper is later overlooked when the radiation is discovered.)

1950

Jan Oort of the Netherlands, building on the previous work of Estonian Ernst Opik, proposes that a vast sphere of comets surrounds the solar system. It becomes known as the Oort Cloud.

1951

Gerard Kuiper argues that there is a belt of sun-orbiting comets at the edge of the solar system, distinct from the Oort Cloud.

1957

The Soviet Union launches *Sputnik,* the world's first artificial satellite.

1960

Frank Drake begins the first electronic search for extraterrestrial intelligence using the National Radio Astronomy Observatory in Green Bank, West Virginia.

1961

Soviet cosmonaut Yuri Gagarin becomes the first human in space during a 108-minute voyage aboard *Vostok I.* In a fifteen minute suborbital flight, Alan Shepard becomes the first American in space. President John F. Kennedy announces that the United States will send astronauts to the moon and return them safely within a decade.

1962

Astronaut John Glenn becomes the first American in orbit.

1963

Arno Penzias and Robert Wilson of Bell Labs discover the cosmic microwave background that supports the big bang theory.

1966

The first mechanical probes land on the moon.

1967

Graduate student Jocelyn Bell discovers a pulsar. Suspecting that the rapid radio pulses may be extraterrestrial signals, she and her adviser nickname them "LGM," for "Little Green Men." Further study by Thomas Gold shows that they are emissions from rapidly spinning neutron stars.

1969

Apollo 11 carries Neil Armstrong and Buzz Aldrin to a safe landing on the moon. Armstrong, the first human to set foot on the moon's surface, calls it "a small step for man, a giant leap for mankind."

1970

The Soviet probe *Venera 7* becomes the first to successfully land on the surface of another planet. The high temperature and pressure on Venus soon put the probe out of commission.

1971

A suspected black hole is detected in the binary star system Cygnus X-1.

1972

The National Aeronautics and Space Administration (NASA) launches *Pioneer 10,* which becomes the first probe to explore the outer solar system, returning data and images from the asteroid belt and Jupiter.

1973

Pioneer 11 is launched. It becomes the first probe to send back images of Saturn. Jeremiah Ostriker and James Peebles show that dark matter must be holding the disks of typical galaxies together. NASA launches *Mariner 10,* which returns close-up photographs of Venus and Mercury.

1975

Venera 9 transmits the first image from the surface of Venus, showing a rocky, barren planet.

1976

Viking 1 becomes the first probe to make a soft landing on Mars.

1980

The *Voyager 1* spacecraft sweeps by Saturn, sending back images of the ringed planet. Its partner, *Voyager 2,* makes the same journey a year later and goes on to photograph Uranus and Neptune.

1984

The first of numerous supermassive black holes at the centers of galaxies is indirectly detected. The SETI Institute, dedicated to the search for extraterrestrial intelligence, is founded.

1989

The *Galileo* spacecraft is launched on a mission to explore Jupiter and its moons.

1990

The Hubble Space Telescope is launched. After a troubled start, it becomes the most powerful astronomical instrument ever.

1992

Observations confirm the existence of the Kuiper Belt of comets.

1993

The first confirmed exoplanet is detected by the study of the gravitational wobble of a star.

1994

The Hubble Space Telescope records the spectacular impact of Comet Shoemaker-Levy 9 on the surface of Jupiter.

1996

The Mars *Pathfinder* lifts off to explore the surface of the Red Planet with a robotic rover.

1997

The Cassini-Huygens mission is launched to explore Saturn and its moon Titan. The Hubble Space Telescope records images of supermassive black holes sucking in matter at the center of galaxies.

1998

Two teams discover that the expansion of the universe is accelerating, implying the existence of dark energy.

1999

The Chandra X-Ray Observatory is launched into space. The existence of dark matter is confirmed by the study of gravitational lensing of light from distant galaxies.

2003

Wilkinson Microwave Anisotropy Probe (WMAP) pins down the age of the universe at 13.7 billion years. NASA launches twin robotic explorers on a mission to Mars.

2004

Cassini explores the rings of Saturn, while the *Huygens* probe lands on Titan. Rovers *Opportunity* and *Spirit* find evidence of ancient waters, possibly signaling the potential for life, on the surface of Mars.

2005

The Mars rover *Opportunity* becomes mired in Martian sand but eventually extricates itself.

FOR FURTHER RESEARCH

Books

Anthony F. Aveni, *Stairways to the Stars: Skywatching in Three Great Ancient Cultures.* New York: John Wiley, 1997.

Carole Ann Camp, *American Astronomers: Searchers and Wonderers.* Springfield, NJ: Enslow, 1996.

Guy Consolmagno, *Turn Left at Orion: A Hundred Night Sky Objects to See in a Small Telescope and How to Find Them.* New York: Cambridge University Press, 1995.

Kitty Ferguson, *Prisons of Light: Black Holes.* New York: Cambridge University Press, 1996.

Timothy Ferris, *The Whole Shebang: A State-of-the-Universe(s) Report.* New York: Simon & Schuster, 1997.

John R. Gribbin, *Companion to the Cosmos.* Boston: Little, Brown, 1996.

Trudy Griffin-Pierce, *Earth Is My Mother, Sky Is My Father: Space, Time, and Astronomy in Navajo Sandpainting.* Albuquerque: University of New Mexico Press, 1992.

Michael Hoskin, ed., *The Cambridge Concise History of Astronomy.* Cambridge, UK: Cambridge University Press, 1999.

Karl Hufbauer, *Exploring the Sun: Solar Science Since Galileo.* Baltimore: Johns Hopkins University Press, 1991.

Christopher R. Kitchin, *Journeys to the Ends of the Universe: A Guided Tour of the Beginnings and Endings of Planets, Stars, Galaxies, and the Universe.* Philadelphia: Hilger, 1990.

Leon M. Lederman, *From Quarks to the Cosmos: Tools of Discovery.* New York: Scientific American Library, 1995.

Michael D. Lemonick, *Echo of the Big Bang.* Princeton, NJ: Princeton University Press, 2003.

John S. Lewis, *Worlds Without End: The Exploration of Planets Known and Unknown.* Reading, MA: Perseus Books, 1998.

Benjamin K. Malphrus, *The History of Radio Astronomy and the National Radio Astronomy Observatory: Evolution Toward Big Science.* Malabar, FL: Krieger, 1996.

Olaf Pedersen, *Early Physics and Astronomy: A Historical Introduction.* New York: Cambridge University Press, 1993.

Martin J. Rees, *Just Six Numbers: The Deep Forces That Shape the Universe.* New York: Basic Books, 2000.

Clive Ruggles, *Astronomy in Prehistoric Britain and Ireland.* New Haven, CT: Yale University Press, 1999.

Carl Sagan, *Pale Blue Dot: A Vision of the Human Future in Space.* New York: Random House, 1994.

Archibald Henry Sayce, *Astronomy and Astrology of the Babylonians.* San Diego: Wizards Bookshelf, 1981.

Noel M. Swerdlow, *The Babylonian Theory of the Planets.* Princeton, NJ: Princeton University Press, 1998.

René Taton and Curtis Wilson, eds., *Planetary Astronomy from the Renaissance to the Rise of Astrophysics.* New York: Cambridge University Press, 1995.

Hugh Thurston, *Early Astronomy.* New York: Springer, 1994.

Neil de Grasse Tyson and Donald Goldsmith, *Origins; Fourteen Billion Years of Cosmic History.* New York: W.W. Norton, 2004.

Gregory Vogt, *The Solar System: Facts and Exploration.* New York: Twenty-First Century, 1995.

Periodicals

Bruce Balick and Adam Frank, "The Extraordinary Deaths of Ordinary Stars," *Scientific American,* July 2004.

Amy J. Barger, "The Midlife Crisis of the Cosmos," *Scientific American,* January 2005.

Stephen Battersby, "Titan Is 'Shockingly Earth-Like,'" *New Scientist,* January 29, 2005.

David Branch, "Tycho's Mystery Companion," *Nature,* October 28, 2004.

Robert Roy Britt, "Telescope Eyes Possible Alien Asteroid Belt," CNN.com, April 20, 2005. www.cnn.com.

Alicia Chang, "Planet Outside Solar System Is Observed," *USA Today,* May 2, 2005.

Allan Chapman, "The Female Touch" *Astronomy Now,* January 1998.

CNN.com, "Hubble Celebrates 15 Years of Stellar Images," April 25, 2005. www.cnn.com.

Leonard David, "Opportunity Mars Rover Stuck in Sand," April 28, 2005. www.space.com.

Andrea Dobson and Katherine Bracher, "Urania's Heritage: A Historical Introduction to Women in Astronomy," *Mercury,* January/February 1992.

John S. Hey, "Half a Century of Radio Astronomy," *Nature,* July 7, 1983.

Jenny Hogan, "A Whiff of Life on the Red Planet," *New Scientist,* February 16, 2005.

Katy Human, "When the Solar Wind Blows," *Astronomy,* January 2000.

David C. Koo, "Exploring Distant Galaxy Evolution: Highlights with Keck," *Reviews in Modern Astronomy,* 2000.

Richard B. Larson and Volker Bromm, "The First Stars in the Universe," *Scientific American,* September 2004.

Maggie McKee, "Exotic Stars Unveiled at Heart of Milky Way," NewScientist.com, January 13, 2005. www.newscientist.com.

Fulvio Melia, "The Heart of the Milky Way," *American Scientist,* July/August 2000.

Philip Morrison, "Copernicus in His Prime," *American Scientist,* March/April 2003.

Martin Rees, "Is the Search for Alien Life Futile Nonsense?" *New Scientist,* July 12, 2003.

I. Neill Reid, "Astronomy: Weighing the Baby," *Nature,* January 20, 2005.

Eric P. Rubenstein, "Superflares and Giant Planets," *American Scientist,* January/February 2001.

Michael Schirber, "Snowball Fight in Saturn's Rings," Space.com, December 20, 2004. www.space.com.

Peter de Selding, "Huygens Probe Sheds New Light on Titan," Space.com, January 21, 2005. www.space.com.

J. Michael Shull, "Astronomy: Hot Pursuit of Missing Matter," *Nature,* February 3, 2005.

Web Sites

Amateur Astronomers Association of New York, www.aaa.org. The Amateur Astronomers Association of New York was founded in 1927. It operates a wide range of public and in-house educational programs in astronomy. The site includes numerous articles and images produced by members.

American Astronomical Society, www.aas.org. The site of America's oldest professional organization representing astronomers. It includes astronomical images, a section on women in the profession, and a section on historical astronomy.

Astronomical Society of the Pacific, www.astrosociety.org. The Astronomical Society of the Pacific was founded in 1889 by a group of northern California professional and amateur astronomers after joining together to view a rare total solar eclipse. It claims to be the largest general astronomy society in the world. Its site includes news, publications, and resources.

Astrophysical.org, www.astrophysical.org. A site run by the Scientia Astrophysical Organization, an association of professional astronomers, astrophysicists, physicists, amateur astronomers, and philosophers interested in astronomy. Its purpose is to popularize astronomy, astrophysics, astrometry, and celestial mechanics. The Web site includes both images and explanations of astronomy concepts.

Best of the Hubble Telescope, www.seds.org. A collection of spectacular images produced by the orbiting Hubble Space

Telescope. The images are copyrighted but may be viewed at the site.

Cassini-Huygens Mission to Saturn and Titan, www.saturn. jpl.nasa.gov. A site operated by the Jet Propulsion Laboratory to inform the public about the latest exploration of the ringed planet. Cassini-Huygens is an international collaboration between three space agencies. Seventeen nations contributed to building the spacecraft. The *Cassini* orbiter was built and managed by the National Aeronautics and Space Administration Jet Propulsion Laboratory. The *Huygens* probe was built by the European Space Agency.

Chandra X-Ray Center, www.xrtpub.harvard.edu. A Web site about the Chandra X-Ray Observatory. Since its launch in 1999, Chandra has been observing X-rays from high-energy regions of the universe, such as the remnants of exploded stars. Many spectacular images are available on the site.

History of Women in Astronomy, www.cannon.sfsu.edu. An extensive set of capsule biographies of women astronomers throughout history.

Infrared Space Observatory, www.iso.vilspa.esa.es. The Infrared Space Observatory is a European Space Agency mission to image astronomical objects in infrared wavelengths, with the participation of Japan and the United States. Site includes images and information.

National Aeronautics and Space Administration, www.nasa.gov. The National Aeronautics and Space Administration, better known as NASA, conducts U.S. space exploration and research. Its site includes information on the history of space exploration as well as current missions.

National Astronomy and Ionosphere Center, www.naic.edu. This Web site is hosted by the Arecibo Observatory, the world's largest single-dish radio telescope. The National Astronomy and Ionosphere Center's dish enables research in the areas of astronomy, planetary studies, and space and atmospheric sciences. The site includes information and images for the public.

INDEX